Axelle Amon, Marc Lefranc
Nonlinear Dynamics

Also of Interest

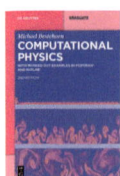

Computational Physics
With Worked Out Examples in FORTRAN® and MATLAB®
Michael Bestehorn, 2023
ISBN 978-3-11-078236-3, e-ISBN 978-3-11-078252-3

Linear Algebra
A Minimal Polynomial Approach to Eigen Theory
Fernando Barrera-Mora, 2023
ISBN 978-3-11-113589-2, e-ISBN 978-3-11-113591-5

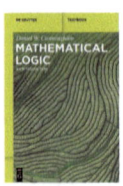

Mathematical Logic
An Introduction
Daniel Cunningham, 2023
ISBN 978-3-11-078201-1, e-ISBN 978-3-11-078219-6

Abstract Algebra
An Introduction with Applications
Derek J. S. Robinson, 2022
ISBN 978-3-11-068610-4, e-ISBN 978-3-11-069116-0

Numerical Methods with Python
for the Sciences
William Miles, 2023
ISBN 978-3-11-077645-4, e-ISBN 978-3-11-077664-5

Axelle Amon, Marc Lefranc

Nonlinear Dynamics

—

DE GRUYTER

Mathematics Subject Classification 2020
Primary: 34-01, 37-01

Authors

Assoc. Prof. Axelle Amon
Université de Rennes
Institut de Physique de Rennes
263 Avenue du Général Leclerc
35042 Rennes
France
axelle.amon@univ-rennes.fr

Prof. Marc Lefranc
Université de Lille
Laboratoire de Physique des Lasers, Atomes,
Molécules
Bat P5
59655 Villeneuve D'Ascq Cedex
France
marc.lefranc@univ-lille.fr

Illustrations by Marion Erpelding

ISBN 978-3-11-067786-7
e-ISBN (PDF) 978-3-11-067787-4
e-ISBN (EPUB) 978-3-11-067807-9

Library of Congress Control Number: 2023935966

Bibliographic information published by the Deutsche Nationalbibliothek
The Deutsche Nationalbibliothek lists this publication in the Deutsche Nationalbibliografie;
detailed bibliographic data are available on the Internet at http://dnb.dnb.de.

Foreword

This textbook is the fruit of master's courses that the authors have taught at their respective universities. It is intended for science students with a basic training in mathematics, especially linear algebra and differential equations, as well as in physics, mainly in classical mechanics. Since our students had very different backgrounds, ranging from earth sciences and biology to fundamental physics, our aim has been to write a book that can be read at different levels and that will satisfy both those who want to learn the basics of nonlinear dynamics and those who want to tackle more advanced mathematical notions. Because geometry is essential to understand the concepts we will discuss, a physicist turned illustrator joined this project, bringing a beautiful point of view on the phenomena studied.

Why study nonlinear dynamics, a rather mathematical subject, when one is interested in applied sciences? The main reason is that almost everything around us is nonlinear. The mathematician and physicist Stanislaw Ulam is supposed to have said *"Using a term like nonlinear science is like referring to the bulk of zoology as the study of non-elephant animals"*. Yet, a very large part of what we have been taught was about linear phenomena. Indeed, these systems are easy to manipulate and can be solved analytically. However, in reality, they are often simplifications of the real problem and only valid in a limited parameter range.

Nonlinear problems are difficult: they generally have no analytical solution, their behavior is rarely intuitive, and they can give rise to very complex behavior that does not arise from stochastic processes. Even though they obey strict physical laws, they can appear erratic and uncontrollable. An interesting example is smoke rising above a candle: just above the flame, the smoke rises steadily in a laminar flow but quickly turns into complex vortices. In both regimes, the laws of fluid mechanics, thermodynamics, and chemistry govern the dynamics, but the two behaviors are completely different.

Does this mean that we are clueless in the face of such problems and that any prediction is out of reach? No, and it is the purpose of this book to give mathematical tools to tackle this kind of problem. Of course, if we know the laws governing the system, then an approach could be using numerical simulations to compute the time courses of the state variables. However, this is a black box giving us little insight about the key ingredients generating the behavior observed.

As we will see, a qualitative theory of dynamical systems can be elaborated, which casts light on the universal mechanisms of dynamical behavior. As often in science, the point is to delineate precisely what we want to know and what we do not need to know: do we want to determine every aspect of the system evolution with time, or do we restrict ourselves to the asymptotic behavior? Do we want to determine what happens for every possible set of control parameters, or do we restrict ourselves to mapping the parameter space into regions where the behavior is qualitatively similar? We will see that the proper way to address this type of problem is to have a geometrical approach and

https://doi.org/10.1515/9783110677874-201

that by looking for the right structures we can understand how the dynamics of these systems are organized.

The textbook is organized as follows. The first chapter is a general introduction that presents the main concepts necessary to describe nonlinear dynamical systems. In particular, the phase space and invariant sets are introduced. The second chapter deals with the linear stability analysis of fixed points. In the third chapter, we discuss bifurcations and how the existence or stability of fixed points is modified when parameters are varied. The fourth chapter is devoted to oscillatory behavior: how it emerges and how to study its stability, and also how it responds to external driving. The fifth chapter of the book is devoted to complex behaviors that need three dimensions to unfold, quasi-periodicity and deterministic chaos. In the sixth and last chapter, we discuss how to characterize chaos, and we study in detail a universal route to chaos, the period-doubling cascade.

We feel indebted to all those that helped us to build the knowledge that we share in this book. M. L. would particularly like to thank Pierre Glorieux and Robert Gilmore, who have been most inspiring to him and guided his bifurcations into fascinating aspects of nonlinear dynamics and chaos. For all the exciting discussions we had with them, we also thank all the smart members of the "Dynamics of complex systems" team at the Laboratoire de Physique des Lasers, Atomes, Molécules of the University of Lille. We are also very grateful to François Maucourant, Elias Charnay, and Yaniss Rabahi for their careful reading of the manuscript.

We thank our loved ones for their constant support and their patience.

Contents

1 Introduction to dynamical systems

In this first chapter, we provide the reader with the essential mathematical tools for the study of dynamical systems. Mostly, those systems describe a dynamics continuous with time and are defined by sets of ordinary differential equations. However, we will show that the study of such systems leads naturally to discrete-time dynamical systems defined by mappings of state spaces into themselves, and we will also introduce the basic tools to study those recurrent systems.

The chapter is organized as follows: we first come back to the essential notion of determinism, and how it is mathematically translated into differential equations. In this first part, we will introduce a mechanical example (the pendulum) and its variants, which will be used as a guideline throughout the chapter to illustrate all the notions introduced. In Section 1.2, we introduce the geometrical approach that will allow us to understand the dynamics of complex systems. We introduce the concept of *phase space* and discuss how dynamical systems evolve in this space. As geometry and topology are essential to understand dynamics in phase space, in Section 1.3, we focus on how dynamics becomes more and more complex as dimension increases, seizing the opportunity to introduce the concepts that will accompany us throughout the book. Finally, in the last part, we show how the description of dynamical systems can be simplified and their dimensionality can be reduced, introducing discrete-time dynamics and recurrence maps.

1.1 Determinism and the notion of state

The progress of science is mainly based on the ability to predict accurately the future evolution of a system of interest from the information available at the present time. This predictive capacity is based on the concept of determinism. If the laws governing the system under consideration do not change and if there is no uncontrolled external disturbance occurring at an arbitrary time, then our everyday experience tells us that the future of this system should be predictable, at least in the short term.

Formalized mathematically, prediction relies on a model from which the future behavior of the system can be computed from a set of initial conditions. The difficulty is to write such a model for a given problem and, above all, to write a minimal model, i. e., a model that reproduces all the features that we want to describe and that incorporates all the key variables that govern the evolution of the system, but no more. Indeed, the more interacting variables we incorporate in the model and the more parameters to be determined, the less insight the model will provide us into the behavior of the system and into the role of different variables. In this case, having a mathematical model is only valuable when numerical simulations of it can provide us with useful information that is not available otherwise (think about the weather forecast).

An important task is therefore to determine the minimal set of variables necessary to define the state of the system by providing the necessary and sufficient information

https://doi.org/10.1515/9783110677874-001

to constrain and predict its future temporal evolution. To illustrate our point, we introduce a system that will serve as an example throughout the chapter to apply the abstract mathematical notions that will be introduced. This example comes from the field of mechanics because it is the archetypal scientific field where the laws governing the motion of bodies are very well known and where the notion of determinism is best anchored in our daily life experience.

1.1.1 Red thread example

Consider a gravity pendulum. The most obvious variable to describe the state of the system is the angle between the pendulum and the vertical direction. However, this is not sufficient, because we find that starting from the vertical position, different trajectories are observed depending on the initial velocity. However, we also observe that if we always start from approximately the same angle and velocity, then approximately the same behavior is always followed.

This can be formalized as follows. Consider a mass m suspended from a pivot O, subjected to the standard gravity field $\mathbf{g}$ shown in Figure 1.1. The string holding the mass is rigid, massless, and of length l. Using polar coordinates (see Fig. 1.1), the torques associated with the forces are

$$\mathbf{OM} \times \mathbf{T} = \mathbf{0}$$
$$\mathbf{OM} \times (m\mathbf{g}) = -mgl \sin \theta \mathbf{e}_z$$

and the angular momentum:

$$\mathbf{OM} \times (m\mathbf{v}) = ml^2 \dot{\theta} \mathbf{e}_z$$

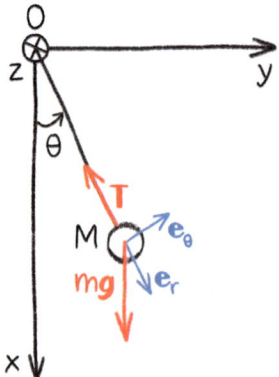

Figure 1.1: Simple gravity pendulum.

Applying the angular momentum conservation and projecting it along $\mathbf{e}_z$, we obtain:

$$\frac{\mathrm{d}}{\mathrm{d}t}(ml^2\dot{\theta}) = -mgl\sin\theta$$

$$\ddot{\theta} + \frac{g}{l}\sin\theta = 0 \tag{1.1}$$

Eq. (1.1) is the pendulum equation of motion. Usually, the problem is studied in the small angle approximation $\sin\theta \simeq \theta$. In this approximation the equation of motion is that of a harmonic oscillator:

$$\ddot{\theta} + \frac{g}{l}\theta = 0$$

which has for general solution:

$$\theta(t) = A\sin\left(\sqrt{\frac{g}{l}}\,t\right) + B\cos\left(\sqrt{\frac{g}{l}}\,t\right), \tag{1.2}$$

where the constants A and B must be determined from the initial conditions. Since there are two unknowns (A and B), two initial conditions are needed to fully solve the problem. The initial angle and the initial angular velocity of the pendulum are generally chosen for this purpose.

Predicting the behavior of a system when the equation describing its dynamics has an analytical solution is straightforward. Nevertheless, the analytical solution (1.2) is valid only in the case of small oscillations. If we want to know all possible trajectories for all possible initial conditions, we have to take into account that the amplitude of the oscillations can be very large and therefore study the original equation of motion (1.1). Such an equation does not have a simple analytical solution. One of the objectives of this book is to provide the general tools to describe the dynamics of any system described by a set of ordinary differential equations.

The gravity pendulum, as well as its variants (damped pendulum and forced pendulum) which will be introduced later, will serve as reference examples to illustrate the various definitions and concepts introduced in this chapter.

1.1.2 Dynamics of state variables

Let us now turn to the general case and consider a system whose state is described by a vector of n state variables $\mathbf{X}(t) = (X_1, X_2, \ldots, X_n)$, which vary smoothly with time. Given the initial condition $\mathbf{X}(t_0) = \mathbf{X}_0$, the system state $\mathbf{X}(t_0 + \tau)$ at a later time $t_0 + \tau$ can be expressed as a Taylor expansion involving the successive time derivatives $\mathbf{X}^{(j)}(t) = \mathrm{d}^j\mathbf{X}(t)/\mathrm{d}t^j$ (if they exist, of course):

$$\mathbf{X}(t_0 + \tau) = \mathbf{X}(t_0) + \tau\dot{\mathbf{X}}(t_0) + \frac{\tau^2}{2}\ddot{\mathbf{X}}(t_0) + \frac{\tau^3}{6}\mathbf{X}^{(3)}(t_0) + \cdots + \frac{\tau^j}{j!}\mathbf{X}^{(j)}(t_0) + \cdots. \tag{1.3}$$

Thus, if we can determine all derivatives $\mathbf{X}^{(j)}(t)$ in terms of $\mathbf{X}(t)$, then expression (1.3) formally solves the problem of predicting future states from the current one. Note that here we set aside the question of whether this problem is well posed over long times. It may be that the end state depends on the initial condition with excessive sensitivity and that uncertainty dramatically increases with time.

Now assume that the laws of motion governing our system specify the time derivatives of the state variables as function of these variables, i. e.,

$$
\begin{cases}
\frac{dX_1}{dt} = F_1(X_1, X_2, \ldots, X_n), \\
\frac{dX_2}{dt} = F_2(X_1, X_2, \ldots, X_n), \\
\quad \vdots \\
\frac{dX_n}{dt} = F_n(X_1, X_2, \ldots, X_n).
\end{cases}
\tag{1.4}
$$

The quantities designated by $X_1, X_2, \ldots, X_n$ can be, for example:
- positions and velocities of bodies submitted to mutual gravitational interaction and conservation of momentum,
- the currents and voltages in an electrical network linked by the electrical laws,
- the number of individuals in interacting populations in an ecosystem,
- the concentrations of reactants in a chemical reaction scheme described by reaction kinetics,
- ...

The system (1.4) is more conveniently written in vector form:

$$
\dot{\mathbf{X}}(t) = \mathbf{F}(\mathbf{X}(t)),
\tag{1.5}
$$

which is the most compact form for a system of first-order ordinary differential equations.

Then we can readily compute the second-order derivative using the chain rule:

$$
\ddot{\mathbf{X}}(t) = \frac{d}{dt}\dot{\mathbf{X}}(t) = \frac{d}{dt}\mathbf{F}(\mathbf{X}(t)) = \frac{\partial \mathbf{F}(\mathbf{X})}{\partial \mathbf{X}} \frac{d\mathbf{X}(t)}{dt} = \frac{\partial \mathbf{F}(\mathbf{X})}{\partial \mathbf{X}} \mathbf{F}(\mathbf{X}),
$$

where $\partial \mathbf{F}(\mathbf{X})/\partial \mathbf{X}$ is the so-called Jacobian matrix

$$
\left(\frac{\partial \mathbf{F}(\mathbf{X})}{\partial \mathbf{X}} \right)_{ij} = \frac{\partial F_i(\mathbf{X})}{\partial X_j}.
$$

Using the same trick, all higher-order derivatives $\mathbf{X}^{(j)}$ can be computed and inserted into the Taylor expansion (1.3), effectively determining the future state $\mathbf{X}(t + \tau)$ from the sole knowledge of the current state $\mathbf{X}(t)$. Therefore, when a dynamical system is defined by a system of first-order differential equations such as (1.5), $\mathbf{X}(t)$ is a faithful state vector as the initial condition $\mathbf{X}(t_0) = \mathbf{X}_0$ constrains entirely the future trajectory of the system. This is the essence of the unicity theorem, which will be stated below.

Note that numerical algorithms for solving differential equations, such as the famous Runge–Kutta methods (Press et al., 2007), basically rely on the Taylor expansion (1.3), combining evaluations of X at different times to recreate the expansion. For example, let us consider the so-called midpoint rule:

$$X_1 = X(t) + \frac{\tau}{2} F(X(t), t),$$

$$X(t + \tau) \approx X(t) + \tau F\left(X_1, t + \frac{\tau}{2}\right).$$

It can be verified by direct substitution that the formula obtained agrees with expansion (1.3) up to order 2.

If the system is instead governed by an ordinary differential equation expressing the kth-order derivative as a function of the variable and of the $k - 1$ lowest-order ones,

$$X^{(k)} = L(X, \dot{X}, \ddot{X}, \ldots, X^{(k-1)}), \tag{1.6}$$

then we see that all high-order derivatives can again be computed recursively with $X^{(j+1)} = dX^{(j)}/dt$, providing us everything we need to compute expansion (1.3) given the vector

$$Y = (X, \dot{X}, \ddot{X}, \ldots, X^{(k-1)}), \tag{1.7}$$

which thus contains the information needed to relate $X(t + \tau)$ to $X(t)$ and hence is a genuine state vector.

Seeing that we have then

$$\dot{Y} = (\overbrace{\dot{X}, \ddot{X}, \ldots, X^{(k-1)}}^{(Y_2, \ldots, Y_k)}, L(Y)) = F(Y),$$

we see that (1.5) is the most general formulation of a temporal dynamical system that is invariant by time translation, and we will now assume that such systems, termed *autonomous*, can always be given in this form.

What now if the differential equation

$$\dot{X} = F(X(t), t)$$

depends on time, expressing that our system is forced from the outside? The system is then said to be *nonautonomous*. Then t must be considered as a state variable, since we need to know its value to predict the future. Including t in a state vector $Y = (X, t)$, we eventually get $\dot{Y} = G(Y) = (F(X(t), t), 1)$ using the obvious $\dot{t} = 1$. This view is most useful when F depends periodically on t, giving rise to a periodic coordinate of the state space.

It is now time to state an important mathematical result concerning systems of the form of Eq. (1.5). The Cauchy–Lipschitz theorem guarantees the existence and unicity of a solution to Eq. (1.5). Indeed, for the initial value problem

$$\begin{cases} \frac{d\mathbf{X}}{dt} = \mathbf{F}(\mathbf{X}, t), \\ \mathbf{X}(0) = \mathbf{X_0}, \end{cases}$$

with $\mathbf{X} \in \mathbb{R}^n$ and all F_i and their partial derivatives $\frac{\partial F_i}{\partial X_j}$ continuous in an open ensemble Ω such that $\mathbf{X_0} \in \Omega$, there is a unique solution $\mathbf{X}(t)$ for $t \in]t_-(\mathbf{X_0}), t_+(\mathbf{X_0})[$, where $t_-(\mathbf{X_0}) < 0 < t_+(\mathbf{X_0})$. This interval can be infinite or finite, as some solutions can blow up to infinity in finite time.

1.1.3 Application to the pendulum

Let us apply the general framework introduced in the previous section to the pendulum system. Defining $X_1 = \theta$ and $X_2 = \dot{\theta}$, Eq. (1.1) can be rewritten as

$$\dot{X}_1 = \dot{\theta} = X_2,$$
$$\dot{X}_2 = \ddot{\theta} = -\frac{g}{l} \sin \theta = -\frac{g}{l} \sin X_1, \tag{1.8}$$

which is an autonomous system of the form $\frac{d\mathbf{X}}{dt} = \mathbf{F}(\mathbf{X})$ with

$$\mathbf{X} = \begin{pmatrix} \theta \\ \dot{\theta} \end{pmatrix} \quad \text{and} \quad \mathbf{F} : \begin{pmatrix} X_1 \\ X_2 \end{pmatrix} \mapsto \begin{pmatrix} X_2 \\ -\frac{g}{l} \sin X_1 \end{pmatrix}.$$

In this system, the energy is conserved over time, but in real systems, there is always some dissipation of the energy. We will see in Section 1.2.5 that conservative and dissipative systems have fundamental differences in their dynamics. Consequently, it will be useful for later on to have an example of a dissipative system. In the case of the pendulum, dissipation can be caused by a viscous torque, which we describe in the equation of motion by a damping term proportional to the angular velocity and opposing motion ($\ddot{\theta} = -\gamma\dot{\theta}$), leading to the new equation of motion

$$\ddot{\theta} = -\gamma\dot{\theta} - \frac{g}{l} \sin \theta,$$

which can be rewritten as

$$\begin{cases} \dot{X}_1 = X_2, \\ \dot{X}_2 = -\frac{g}{l} \sin X_1 - \gamma X_2. \end{cases} \tag{1.9}$$

We know that without any energy input, such a dissipative system will always converge toward an asymptotic state in which it is at rest. To maintain the motion of a dampened pendulum, it must be driven, usually by exerting a periodic torque $M\sin(\omega t)$ at a fixed frequency ω, which leads to the equation of motion

$$\ddot{\theta} + \gamma\dot{\theta} + \frac{g}{l}\sin\theta = M\sin(\omega t).$$

With $X_1 = \theta$ and $X_2 = \dot{\theta}$, we obtain the system

$$\begin{cases} \dot{X}_1 = X_2, \\ \dot{X}_2 = -\frac{g}{l}\sin X_1 - \gamma X_2 + M\sin(\omega t), \end{cases}$$

which is a nonautonomous system. The variable t appears explicitly in the vector field. As discussed in Section 1.1, introducing a new variable $X_3 = \omega t$ leads us to

$$\begin{cases} \dot{X}_1 = X_2, \\ \dot{X}_2 = -\frac{g}{l}\sin X_1 - \gamma X_2 + M\sin X_3, \\ \dot{X}_3 = \omega, \end{cases} \tag{1.10}$$

which is an autonomous system of dimension 3.

Those nonlinear systems of differential equations can be integrated using numerical algorithms such as the Runge–Kutta method; an example of Python script achieving this task is provided in the section of exercises. An example of such a numerical integration for the pendulum equations is shown in Fig. 1.2. In particular, Fig. 1.2a displays the amplitude θ and angular velocity $\dot{\theta}$ in the case of the nonlinear conservative pendulum. Oscillatory behavior of constant amplitude at the natural frequency of the system is observed as expected. In the damped case (Fig. 1.2b) the amplitude of the oscillations decreases exponentially with time. In the case of the driven pendulum, depending of the values of the parameters, either regular behavior at the imposed frequency (Fig. 1.2c) or complex dynamics (Fig. 1.2d) can be observed. Note that the dynamics shown in Fig. 1.2d corresponds to an asymptotic regime and that due to 2π-periodicity, the graph has periodic limit conditions along the y-axis.

1.2 Geometrical description of the dynamics

1.2.1 Velocity vector field

To build a theory as general as possible, we need to build a description of the dynamics that is independent of the precise form of Eqs. (1.5) and that allows us to analyze the succession of states visited by the system over a possibly infinite amount of time. To

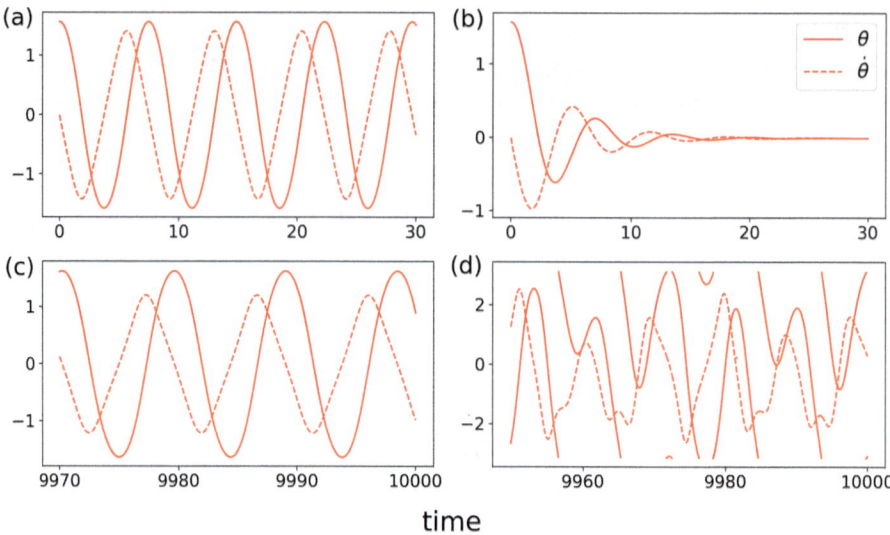

Figure 1.2: Numerical integration of (a) conservative pendulum (Eq. (1.8)), oscillating without damping at its natural frequency; (b) damped pendulum (Eq. (1.9)), relaxing toward equilibrium; (c)–(d) driven pendulum (Eq. (1.9)) with different responses for different parameter values: (c) sustained oscillations at the imposed frequency ω; (d) chaotic behavior. Solid (resp., dashed) line: amplitude θ (resp., angular velocity $\dot{\theta}$) as a function of time. Note that θ is 2π-periodic.

achieve this, the key insight is that the differential equation

$$\dot{\mathbf{X}}(t) = \mathbf{F}(\mathbf{X}(t)) \tag{1.11}$$

has a natural interpretation as specifying the velocity vector $\dot{\mathbf{X}}(t)$ of a point moving in an abstract space whose coordinates are the state variables $\mathbf{X}(t)$. The point is called the representative point of the system, and the space is called the state space or more usually the *phase space*. The dimension of the phase space is the number of state variables and thus can be arbitrarily large, although we will see that interesting dynamics arises already for a few degrees of freedom. In the following, we will typically denote the phase space by $\mathcal{S}$.

To each possible state of the system, there corresponds a point in the phase space. The velocity vector field (1.11) at that point indicates in which direction and how fast the system is moving in the phase space. This is similar to an orienteering race, where signs placed along the (invisible) track tell you in which direction to proceed. There is also a clear analogy with stationary hydrodynamics, where particles of fluid move across a physical space along flow lines that are tangent everywhere to the fluid velocity.

Application to the pendulum – Let us look at this vector field in the case of the conservative pendulum (Eqs. (1.8)). The state space has dimension 2, which will allow us to draw it in a plane. Moreover, θ is 2π-periodic, so that we only need to draw the vector field for $\theta \in [-\pi, \pi]$. Let us compute $\mathbf{F}$ in a few specific points (the corresponding vectors are drawn as blue solid arrows in Figure 1.3):

$$\mathbf{F}\begin{pmatrix} \pi/2 \\ 0 \end{pmatrix} = \begin{pmatrix} 0 \\ -g/l \end{pmatrix} \quad \mathbf{F}\begin{pmatrix} -\pi/2 \\ 0 \end{pmatrix} = \begin{pmatrix} 0 \\ g/l \end{pmatrix},$$

$$\mathbf{F}\begin{pmatrix} 0 \\ \dot{\theta}_0 \end{pmatrix} = \begin{pmatrix} \dot{\theta}_0 \\ 0 \end{pmatrix} \quad \mathbf{F}\begin{pmatrix} \pm\pi \\ \dot{\theta}_0 \end{pmatrix} = \begin{pmatrix} \dot{\theta}_0 \\ 0 \end{pmatrix}.$$

(1.12)

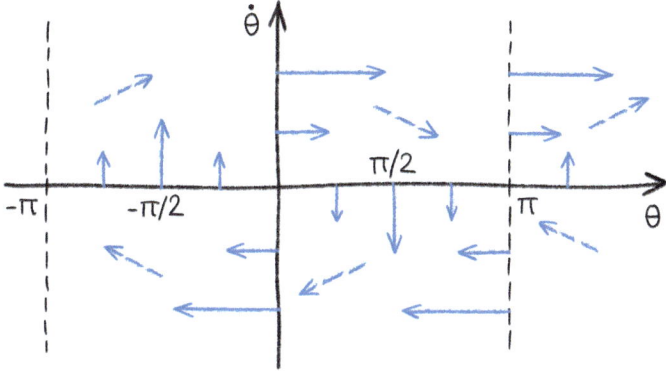

Figure 1.3: Representation of the vector field defined by (1.8). The blue arrows represent the vector field **F** at the considered points. Note that the problem is periodic along the horizontal axis, so that we can restrict the phase space between $-\pi$ and π along this direction.

Calculating the velocity vectors in many places is tedious, although computers do it easily for us, and only needed if we want to determine specific trajectories precisely. To get a global picture, it is useful to display the general direction of the vector field in different areas of the phase space. As the functions F_1 and F_2 constituting the system (1.12) are both odd functions, the orientations of the vector field in each quadrant of the phase space delimited by the axes are easy to determine and are drawn with dashed blue arrows in Figure 1.3.

Let us now plot the vector field in the case of the *damped* pendulum. We can again compute the values of its components at typical points:

$$\begin{pmatrix} \theta_0 \\ 0 \end{pmatrix} \mapsto \begin{pmatrix} 0 \\ -g/l \sin \theta_0 \end{pmatrix} \quad \begin{pmatrix} 0 \\ \dot{\theta}_0 \end{pmatrix} \mapsto \begin{pmatrix} \dot{\theta}_0 \\ -\gamma\dot{\theta}_0 \end{pmatrix}.$$

The general directions of the vector field are now more difficult to draw as the four possible directions of $\mathbf{F}$ do not correspond to the four quadrants of the plane. Still, we can draw the curve $y = f(\theta) = -\frac{g}{\gamma l} \sin \theta$, which delimits the change of sign of $\ddot{\theta}$. Together with the y-axis, it separates the areas corresponding to different orientations of $\mathbf{F}$ (Fig. 1.4). Such curves are called nullclines and will be discussed in Section 1.3.2.

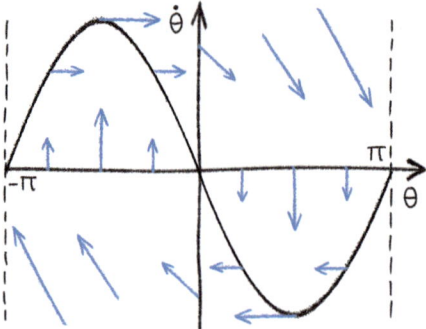

Figure 1.4: Vector field corresponding to Eqs. (1.9).

1.2.2 Flow and trajectory

Following the vector field $\mathbf{F}(\mathbf{X})$, states in the phase space are sent to other states, like a hydrodynamic flow sends fluid particles at a given location to other locations. The application that maps the state $\mathbf{X}(t_0)$ at time t_0 to state $\mathbf{X}(t_0 + \tau)$ at time $t_0 + \tau$ is called the *flow* ϕ_τ associated with the vector field $\mathbf{F}(\mathbf{X})$:

$$\phi_\tau(\mathbf{X}(t_0)) = \mathbf{X}(t_0 + \tau). \tag{1.13}$$

It is often more revealing to consider how the flow ϕ_τ sends finite regions of the phase space to other regions than merely studying its action on isolated states. This approach will be particularly useful when we will study chaotic behavior, as we shall see in Section 5.3.3.

Given $\mathbf{X}_0 = \mathbf{X}(t_0)$, the set of locations $\gamma(\mathbf{X}_0) = \{\mathbf{X}(t); t \in \mathbb{R}\} = \{\phi_\tau(\mathbf{X}_0); \tau \in \mathbb{R}\}$ successively visited in the phase space by the representative point of the system is called the *orbit* of $\mathbf{X}_0$ and represents the trajectory of the system in the phase space, like a streamline in a stationary fluid flow in hydrodynamics. Such an orbit is represented as a red solid line in Fig. 1.5. The trajectory is everywhere tangent to the velocity vector field $\dot{\mathbf{X}}(t)$ by the definition of the latter (blue arrow in Fig. 1.5).

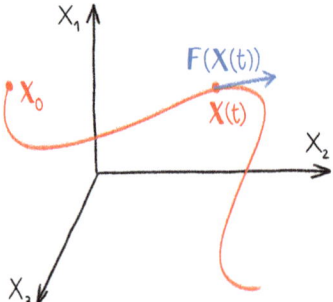

Figure 1.5: Representative point $\mathbf{X}(t)$ and its trajectory in phase space, velocity vector field $\mathbf{F}(\mathbf{X})$ at $\mathbf{X}(t)$.

A geometric formulation of the unicity theorem is that one and only one trajectory can go through a given point in the phase space if the vector field is not zero at that point. In other words, two trajectories cannot cross transversely at a point. Figure 1.6 illustrates this fundamental property: since each of the two trajectories would be tangent to a velocity vector, there would be two velocity vectors at the same point of the phase space, contradicting the fact that (1.5) determines a unique $\mathbf{F}(\mathbf{X})$ at any given state. This is an important fact that constrains trajectories very much.

Figure 1.6: Two trajectories cannot intersect in the phase space.

Application to the pendulum – Let us revisit the phase portraits in Figs. 1.3 and 1.4 to deduce typical trajectories in the phase space. In both cases, the shape of the trajectories can be deduced from the organization of velocity vectors.

For the conservative pendulum, we see in Fig. 1.3 that around the origin, the vectors rotate along ellipses. This is in agreement with the analytical solution in the small angle approximation (see Section 1.1.1). These solutions, which correspond to oscillations, are associated with closed orbits around the origin (see Fig. 1.7). Far from the origin, for large initial $\dot{\theta}_0$, we obtain trajectories that never cross the θ-axis, indicating that the angular velocity never cancels during the motion. The pendulum does not swing back and forth around the vertical axis but rotates continuously around the pivot.

Two trajectories plotted in orange in Fig. 1.7 display a very strange behavior in the vicinity of the points $(\pi, 0)$ and $(-\pi, 0)$, which are in fact the same point since the angle

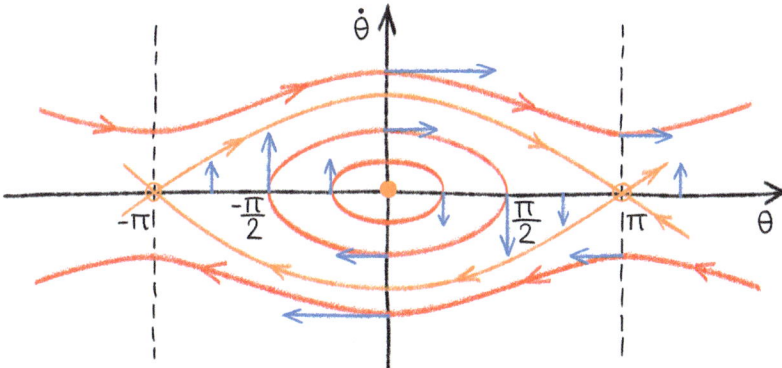

Figure 1.7: Trajectories of a pendulum in the phase space.

coordinate is 2π-periodic. A pair of orange trajectories converge toward those points, whereas another pair moves away from it. We will see in the following that those trajectories have a very specific role in the organization of the dynamics in the phase space. In particular, they separate the two types of trajectories, oscillations and rotations. A trajectory that delimits two different behaviors is called a *separatrix*.

The phase portrait of the damped pendulum (Fig. 1.4) shows few differences from that of the conservative one. We mainly observe that the vector field along the y-axis is not horizontal but points to the x-axis. This means in particular that rotations cannot survive indefinitely as the distance to origin decreases at each revolution. In fact, whatever the initial condition and for a not too strong damping, all trajectories end up spiraling toward the point $(0,0)$ (or any point $(2n\pi, 0)$, since θ is 2π-periodic), as shown in Fig. 1.8. Such a trajectory is the representation in the phase space of the damped temporal dynamics shown in Fig. 1.2.

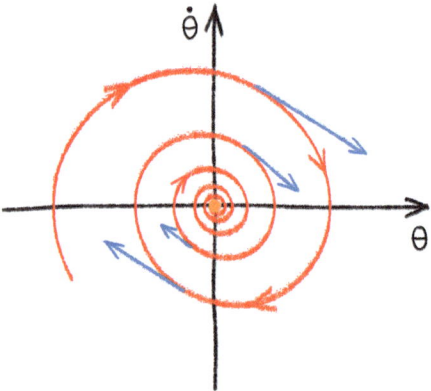

Figure 1.8: An example of trajectory of a damped pendulum in the phase space in the case where the damping is weak (i. e., where $y^2/4 < g/l$; see Example 2.1.6.a in the next chapter). For large damping, the system goes directly to the origin without spiraling. Those two cases correspond respectively to the underdamped and overdamped regimes.

1.2.3 Fixed points

Several points play a particular role in the examples discussed in the previous part. In the conservative case, we underlined the peculiar behavior of the separatrices around the points $(\pm\pi, 0)$. In the damped case, all the trajectories converge toward the point $(0,0)$. This does not contradict the no-crossing theorem because the velocity vector vanishes in these points and thus has no specific orientation: $\mathbf{F(X)} = 0$. Such points can therefore belong to different trajectories, which in fact will slow down progressively as they approach the points, taking an infinite amount of time to reach them. This holds both for the attracting origin in the damped case and for the intersection of the separatrices shown in orange in Fig. 1.7.

The points $\mathbf{X}^*$ where the velocity vector $\mathbf{F}(\mathbf{X}^*) = 0$ play a very important role in the dynamics. Such points, called *fixed points*, correspond to a stationary dynamics, as all state variables remain constant in time. In the phase space, the fixed points are organizing centers which structure the flow. Not only does the dynamics generally slow down in their neighborhood, but also several trajectories can converge to them in infinite time, as if they were crossing each other.

A particularly important property of fixed points is that they are invariant under the action of the flow:

$$\phi_\tau(\mathbf{X}^*) = \mathbf{X}^* \quad \forall \tau \in \mathbb{R}. \tag{1.14}$$

Fixed points are the simplest examples of *invariant sets*, which are essential features of a dynamical system. An invariant set Λ is such that any trajectory starting in it remains in it, and thus it satisfies

$$\forall \tau \in \mathbb{R}, \quad \phi_\tau(\Lambda) = \Lambda. \tag{1.15}$$

Invariant sets are important because a trajectory cannot cross an invariant set transversally; it can only remain within it indefinitely. Thus invariant sets, depending on their dimensionality, can behave as barriers in the phase space, dividing it into disconnected regions. Important examples of invariant sets that we will encounter are fixed points, limit cycles associated with oscillations (Chapter 4), invariant tori associated with quasi-periodic behavior, and strange attractors associated with chaos (Chapter 5).

1.2.4 Boundedness, fixed points, and recurrences

Fixed points are not exotic but rather a general feature of dynamical systems. Indeed, it is often the case that for physical reasons, the system cannot escape to infinity and remains inside a bounded region $\mathcal{B}$ of the phase space. This means that any trajectory originating from this bounded region remains in it forever. Translating this property in terms of the flow (1.13), this implies that $\phi_\tau(\mathcal{B}) \subset \mathcal{B}$ for all $\tau \geq 0$.

If the region $\mathcal{B}$ is homeomorphic to a n-dimensional ball,[1] then the Brouwer fixed-point theorem can be invoked to show that $\mathcal{B}$ encloses at least one fixed point. This theorem is a generalization of the intermediate-value theorem in one dimension. The latter states that any continuous function on an interval $[a, b]$ takes any value in $[f(a), f(b)]$ at some point in the interval. A consequence is that if $[f(a), f(b)] \subset [a, b]$, then by the intermediate-value theorem applied to $g(x) = f(x) - x$ there is $c \in [a, b]$ such that $f(c) = c$.

---

[1] I. e., there is a 1-1 mapping from the region to an n-dimensional ball, preserving topological properties such as having no interior hole.

Similarly, the Brouwer fixed-point theorem states that any continuous function from a closed ball (or any region homeomorphic to a closed ball) into itself has a fixed point. An important consequence is that any dynamical system evolving in a phase space that is bounded and has no interior hole, which is the case for most physical systems, has a fixed point.

However, we will see that fixed points do not always represent the asymptotic behavior of a system, as they may fail to attract all trajectories in their neighborhood. In this case the dynamics evolves on more complex invariant sets.

When the system does not converge to a stable fixed point, we can show that there always will be points whose neighborhood is visited again and again as time flows. The points to which the system returns arbitrarily close and infinitely many times are called *recurrent*. Such points are important because we focus on the asymptotic dynamics of a system. Consequently, we are not interested in states whose neighborhood will be visited a finite number of times and then never again.

To understand why such recurrent states always exist, let us consider the trajectory starting from a given initial condition with a ball of diameter ε surrounding the system state. As the trajectory unfolds, this ball drills a tube of transverse diameter ε and of steadily increasing length in the phase space, since the velocity is bounded from zero for a trajectory not converging to a fixed point. Thus the volume of this tube would grow indefinitely with the trajectory, in contradiction with the finiteness of the bounding region, unless the tube intersects itself at some point. At the intersection, there is a point whose trajectory returns closer to it than ε. Since ε can be arbitrarily small, we see that there are points whose orbit returns arbitrarily close to them, and this happens infinitely many times.

To make the definition of a recurrent state more rigorous, we can define a point $\mathbf{Y}$ as a ω-limit point of a point $\mathbf{X}$ if there is a subset $\{\phi_{t_k}(\mathbf{X}); t_0 < t_1 < \cdots < t_k < t_{k+1} < \cdots; t_k \to \infty\}$ of the orbit of $\mathbf{X}$ that converges to $\mathbf{Y}$. In other words, the orbit of $\mathbf{X}$ over an infinite time comes arbitrarily close to $\mathbf{Y}$, arbitrarily many times. The ω-limit set $\omega(\mathbf{X})$ is the set of all ω-limit points of $\mathbf{X}$. A point is *recurrent* if it belongs to its own ω-limit set. The *recurrent set* is the set of all recurrent points. We can define the α-limit set in the same way with sequences of points going backward in time along the orbit of $\mathbf{X}$.

Note that although it is guaranteed that the system will return to an arbitrarily close neighborhood of a recurrent point, the actual return time typically depends on dimensionality and can be astronomically large in high-dimensional state spaces.

1.2.5 Conservative vs. dissipative systems

The dynamics in the phase space is very different depending on if the system is *conservative* or *dissipative*. This can be seen by comparing the dynamics in the phase space of the conservative pendulum and the damped one: in the conservative case, an infinite number of closed orbits are nested into each other, whereas in the damped case, all the trajectories converge toward the point $(0, 0)$ as $t \to \infty$. In this section, we formalize the difference between those two kinds of systems.

1.2.5.a Conservative systems

From a physicist's point of view, a conservative system is a system whose total energy is conserved. From the point of view of dynamical systems, this translates into the fact that the volumes in the phase space are conserved under the action of the flow.

Consider an initial set Ω_0 in the phase space and how it is mapped into another set Ω_t under the action of the flow ϕ_t (Fig. 1.9a).

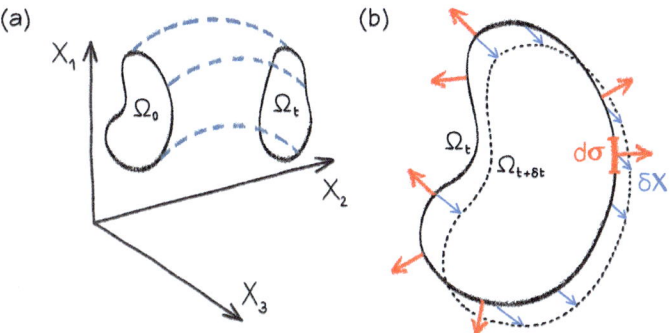

Figure 1.9: (a) Under the action of the flow ϕ_t, a region Ω_0 of the phase space is mapped to another region Ω_t. (b) Infinitesimal displacement of the finite volume Ω_t under the action of the flow between t and $t + \delta t$ (blue arrows). $d\sigma$ is a surface element associated with a normal vector pointing outward the volume (red arrows).

The volume of Ω_t can be expressed as

$$\mathcal{V}(\Omega_t) = \int_{\Omega_t} dX_1 dX_2 \ldots dX_n$$

The change in volume of Ω between t and $t + \delta t$ is given by (Fig. 1.9b)

$$\delta \mathcal{V} = \int_{\partial \Omega} (\delta \mathbf{X}).d\boldsymbol{\sigma},$$

where $\partial \Omega$ is the border of Ω, $\delta \mathbf{X}$ is the change of $\mathbf{X}$ between t and $t + dt$, and $d\boldsymbol{\sigma}$ is the vector associated with a surface element of $\partial \Omega$ (orthogonal to the surface element and with magnitude equal to its area).

As $\delta \mathbf{X} = \mathbf{F}(\mathbf{X})\delta t$,

$$\delta \mathcal{V} = \delta t \int_{\partial \Omega} \mathbf{F}(\mathbf{X}).d\boldsymbol{\sigma},$$

which by the Ostrogradsky theorem can be rewritten as

$$\frac{\delta \mathcal{V}}{\delta t} = \int_{\Omega} (\mathbf{\nabla}.\mathbf{F}) dv,$$

where

$$\mathbf{\nabla}.\mathbf{F} = \frac{\partial F_1}{\partial X_1} + \frac{\partial F_2}{\partial X_2} + \cdots + \frac{\partial F_n}{\partial X_n}$$

is the divergence of the velocity vector field $\mathbf{F}$ and $dv = dX_1 dX_2 \ldots dX_n$.

If the volume is conserved, $\frac{\delta \mathcal{V}}{\delta t} = 0$, which leads to

$$\mathbf{\nabla}.\mathbf{F} = 0. \tag{1.16}$$

This invariance of volumes in the phase space and the associated vanishing of the divergence of the velocity vector field are the mathematical properties defining a conservative system.

Application to the gravity pendulum – In the conservative case, the equations of motion of the pendulum are (Eqs. (1.8))

$$\begin{cases} \dot{X}_1 = X_2, \\ \dot{X}_2 = -\frac{g}{l} \sin X_1. \end{cases}$$

The divergence of the vector field is then

$$\mathbf{\nabla}.\mathbf{F} = \frac{\partial F_1}{\partial X_1} + \frac{\partial F_2}{\partial X_2}$$

$$= \frac{\partial}{\partial X_1}(X_2) + \frac{\partial}{\partial X_2}\left(-\frac{g}{l} \sin X_1\right) = 0.$$

The system is indeed conservative.

1.2.5.b Dissipative systems

In this book, we are mainly interested in dissipative systems. Such systems display asymptotic *volume contraction* in the phase space under the action of the flow, which can be expressed by the fact that on average, we have $\mathbf{\nabla}.\mathbf{F} < 0$.

Application to the damped pendulum – The equations of motion of the damped pendulum are (Eqs. (1.9))

$$\begin{cases} \dot{X}_1 = X_2, \\ \dot{X}_2 = -\frac{g}{l} \sin X_1 - \gamma X_2. \end{cases}$$

Then

$$\mathbf{\nabla.F} = \frac{\partial}{\partial X_1}(X_2) + \frac{\partial}{\partial X_2}\left(-\frac{g}{l}\sin X_1 - \gamma X_2\right)$$
$$= -\gamma.$$

Since $\gamma > 0$ for a damped system, we indeed have $\mathbf{\nabla.F} < 0$.

1.2.6 Attractors

A distinctive property of dissipative systems is the existence of *attractors*, invariant and indecomposable subsets of the phase space that have neighborhoods in which all trajectories converge asymptotically to the attractor.

The existence of attractors is a consequence of volume contraction in the phase space, as any neighborhood of an attractor has its volume shrinking to zero under the action of the flow, thus being reduced to an invariant object. If we consider a trapping region $\mathcal{B}$ such that $\phi_t(\mathcal{B}) \subset \mathcal{B}$ for all $t \geq 0$, then it follows that $\cap_{t>0}\phi_t(\mathcal{B})$ is by definition invariant under the action of the flow, and has zero volume since $\lim_{t\to\infty} V(\phi_t(\mathcal{B})) = 0$.

However, it may be that this invariant subset can be decomposed into disjoint sets $\mathcal{A}_i$, each $\mathcal{A}_i$ having a separate neighborhood in which all trajectories converge to $\mathcal{A}_i$ and such that it cannot be further decomposed into invariant subsets. This implies that all points in an attractor are recurrent; otherwise, the invariant set could be subdivided. Each $\mathcal{A}_i$ is considered as a separate attractor. The region of the phase space where all trajectories converge to $\mathcal{A}_i$ is called the basin of attraction of $\mathcal{A}_i$. When a dynamical system has several attractors, it is said to display *multistability*.

Therefore, if we choose an initial condition on an attractor $\mathcal{A}$, then the system not only remains in $\mathcal{A}$, but it also returns arbitrarily many times and arbitrarily close to the initial condition.

In the case of the damped oscillator (Fig. 1.8), the attractor is the fixed point at $(0,0)$, a zero-dimensional object in a two-dimensional phase space. In the next section, we will discuss more complex attractors (see Sections 1.3.2 and 1.3.3).

1.3 Dynamics and phase space dimension

The type of asymptotic dynamics and invariant sets we can observe depends on the dimension of the phase space in which the system evolves. In this section, we discuss the dynamics of a system constrained to evolve first on a line, then in a plane, and finally free to evolve in a phase space of dimension three or higher.

1.3.1 One-dimensional dynamics

The simplest dynamical systems feature a single state variable and thus are governed by equations such as

$$\dot{x} = f(x) \quad \text{with } x \in \mathbb{R}. \tag{1.17}$$

In some cases, this differential equation may be solved for $x(t)$ by integrating $dt = dx/f(x)$ and inverting the relation obtained. However, the result is generally not worth the effort as there is little value in determining the exact time course of x. Rather, we are interested in the asymptotic dynamics and how the system behaves when transients have died out.

A geometrical approach of the dynamics can answer our questions without going into complicated analytical calculations (Strogatz, 2018). Since only one axis is required to represent the state space, plotting the graph of the function $f(x)$ provides us with all the information we need to capture the asymptotic dynamics.

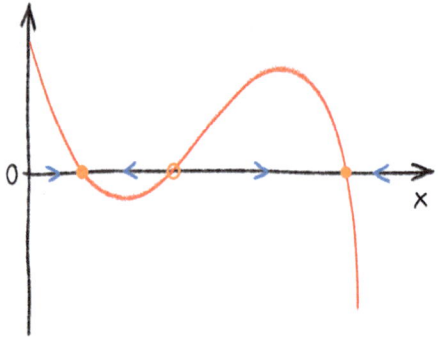

Figure 1.10: A one-dimensional system $\dot{x} = f(x)$ specified by a function $f(x)$ whose zeroes correspond to fixed points. Stable fixed points (with negative slope) are represented by filled circles, and unstable ones (with positive slope) by open circles.

The zeroes of $f(x)$ are the *fixed points* of the system, where the state variable remains constant indefinitely. They represent the simplest example of an invariant set. In Fig. 1.10, the red line is the graph of a function $f(x)$, and orange circles (both filled and open) designate fixed points.

Fixed points separate the phase space into regions where $x(t)$ evolves from left to right ($f(x) > 0$) and regions where $x(t)$ evolves from right to left ($f(x) < 0$) (as shown in Fig. 1.10 with blue arrows). Since the time evolution of $x(t)$ is monotonous until it reaches a fixed point and stops, fixed points are the only possible asymptotic states of the system, unless the latter diverges to infinity. Moreover, trajectories cannot go across fixed points, which form absolute obstacles for the dynamics. This is because 0-dimensional (0D) sets divide 1-dimensional (1D) sets.

However, not every fixed point is involved in the asymptotic dynamics. Indeed, this is only possible if the flow around a fixed point x^* drives the system state toward x^*. This requires that $f(x) > 0$ when $x < x^*$ (on the left of the fixed point) and that $f(x) < 0$ when $x > x^*$ (on the right of the fixed point). Such a fixed point is *stable*. It provides us with the simplest example of an *attractor*, a set that attracts all trajectories in its neighborhood. Using a common convention, stable fixed points will be represented by filled circles (see Fig. 1.10). When f is differentiable, the stability in this sense implies that $f'(x^*) = df(x)/dx|_{x=x^*} < 0$; however, we stress that the sign of $f(x)$ in a neighborhood of x^* suffices to determine the stability.

Conversely, the flow will drive the system away from fixed points x^* such that $f(x) < 0$ (resp., $f(x) > 0$) when $x < x^*$ (resp., $x > x^*$). Such fixed points are *unstable*; they are *repulsors*. When f is differentiable, $f'(x^*) = df(x)/dx|_{x=x^*} > 0$. Unstable fixed points will be represented by open circles (see Fig. 1.10).

These facts about the stability of a fixed point can also be recovered through a differential analysis. Let us study the time evolution of an infinitesimal perturbation δx such that $x = x^* + \delta x$. Equation (1.17) translates into

$$\frac{d}{dt}(x^* + \delta x) = \dot{\delta x} = f(x^* + \delta x) = f(x^*) + f'(x^*)\delta x + \cdots \approx f'(x^*)\delta x.$$

The solution of this differential equation is

$$\delta x(t) = \delta x(0)e^{f'(x^*)t},$$

so that $\lim_{t\to\infty} \delta x(t) = 0$ (resp., ∞) when $f'(x^*) < 0$ (resp., $f'(x^*) > 0$), indicating the stability and instability, respectively.

We can see in Fig. 1.10 that unstable fixed points typically alternate with stable fixed points if $f(x)$ is continuous. This is a simple consequence of the fact that the sign of the derivative of f changes between two consecutive zeroes. Indeed, the crossings are done in opposite directions since the state variable is continuous.

The two unstable fixed points surrounding a stable fixed point x^* delimit the region in which trajectories converge to x^*. This region is called the *basin of attraction* of x^*.

In conclusion, the dynamics of a one-dimensional continuous dynamical system is relatively trivial as it is asymptotically constant. Still, it has allowed us to introduce a few concepts that are relevant in any dynamical system. Moreover, we will see in Chapter 3 that in spite of their simplicity, 1D systems can display complex behavior in how fixed points appear or disappear as $f(x)$ is varied or change their stability, events known as *bifurcations*. Their study will be fruitful since we will see that bifurcations of 1D systems form the backbone of those occurring in higher-dimensional systems.

Note that the topology of the state space matters. Here we have considered systems evolving on the real axis, but it is also possible to consider systems described by a phase variable φ living on a circle S^1. We will encounter such systems in Section 4.3.2 as they naturally describe systems evolving along a periodic orbit.

1.3.2 Two-dimensional dynamics

Two-dimensional systems describe the coupled evolution of two dynamical variables:

$$\dot{x} = f(x,y), \tag{1.18a}$$
$$\dot{y} = g(x,y). \tag{1.18b}$$

In this case the space where the representative point of the system moves is a *phase plane* or a *phase portrait*.

1.3.2.a Fixed points and nullclines
As previously, an important feature of a dynamical system are the fixed points (x^*,y^*) defined by

$$\dot{x} = f(x^*,y^*) = 0, \tag{1.19a}$$
$$\dot{y} = g(x^*,y^*) = 0, \tag{1.19b}$$

which represent invariant states where the state variables remain constant. We will see in Chapter 2 that in 2D, fixed points are not necessarily totally unstable or totally stable: they can be of mixed type, being stable in one direction and unstable in another. Figure 1.11 shows an example of such a fixed point (x_1^*,y_1^*). In the vicinity of this point, we observe that there is a stable direction and an unstable one. The other fixed point (x_2^*,y_2^*) is totally stable.

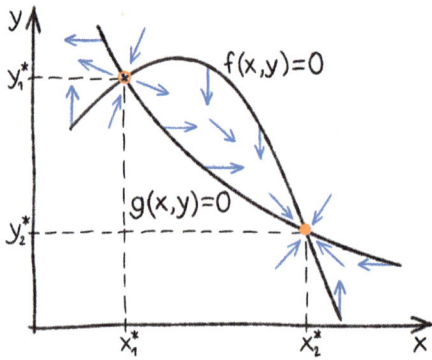

Figure 1.11: Dynamics in a two-dimensional phase plane. The nullclines (where a derivative of one variable or of the other vanishes) are represented by solid dark lines. The fixed points are located at the intersections of the nullclines. Note that when trajectories cross the nullclines, they have either vertical or horizontal velocity, depending on which the time derivative vanishes.

The fact that the vector field vanishes at a fixed point greatly constrains the structure of the vector field in an extended neighborhood. However, fixed points are no longer absolute obstacles for the dynamics since trajectories can revolve around fixed points. A 0D set cannot divide a 2D set, which can only be separated by 1D sets.

As we saw when we plotted the vector fields for the pendulum examples (Section 1.2.1), interesting information can be obtained from the curves where a single time derivative, but not the other, vanishes (i. e., $f(x, y) = 0$ or $g(x, y) = 0$). These curves are called *nullclines* (meaning zero slope). They relatively constrain the structure of the velocity vector field as the latter is either fully vertical (zero x derivative) on the x-nullcline $f(x, y) = 0$ or fully horizontal (zero y derivative) on the y-nullcline $g(x, y) = 0$. In Fig. 1.11 the nullclines are plotted as solid dark lines. The intersections of those curves are the fixed points. When we will study oscillations and how they can appear in a system, we will see that the geometry of nullclines is a key ingredient.

Following a similar approach as in Section 1.3.1, the stability of a two-dimensional fixed point can be studied by studying the time evolution of a deviation from the fixed point. By substituting $x = x^* + \delta x$ and $y = y^* + \delta y$ into (1.18) we have

$$\frac{d}{dt}\delta x = f(x^* + \delta x, y^* + \delta y) = \frac{\partial f}{\partial x}(x^*, y^*)\delta x + \frac{\partial f}{\partial y}(x^*, y^*)\delta y, \qquad (1.20a)$$

$$\frac{d}{dt}\delta y = g(x^* + \delta x, y^* + \delta y) = \frac{\partial g}{\partial x}(x^*, y^*)\delta x + \frac{\partial g}{\partial y}(x^*, y^*)\delta y, \qquad (1.20b)$$

which can be recast in matrix form

$$\frac{d}{dt}\begin{pmatrix} \delta x \\ \delta y \end{pmatrix} = \begin{pmatrix} \frac{\partial f}{\partial x}(x^*, y^*) & \frac{\partial f}{\partial y}(x^*, y^*) \\ \frac{\partial g}{\partial x}(x^*, y^*) & \frac{\partial g}{\partial y}(x^*, y^*) \end{pmatrix}\begin{pmatrix} \delta x \\ \delta y \end{pmatrix}. \qquad (1.21)$$

The matrix appearing in (1.21) is the Jacobian matrix of system (1.18). We will see that the fixed point of the latter is stable when all the eigenvalues of the Jacobian have negative real parts, thus generalizing our findings of Section 1.3.1.

1.3.2.b Stable and unstable manifolds

Nullclines are useful, but they do not organize stringently the phase space. They are not barriers for the dynamics, but only give an indication of the flow direction. As we discussed before, the invariant sets are what really matters to us. Actually, fixed points in the plane are not isolated invariant sets but are dressed with invariant curves that are attached to them. The notion of invariant curves is easier to grasp in the context of a linear system, since they are rooted in a neighborhood of the fixed point that they extend, inside which they are determined by the linear part of the flow.

Consider the following linear system, where for simplicity we assume that the system has been diagonalized (see Chapter 2):

$$\begin{pmatrix} \dot{x} \\ \dot{y} \end{pmatrix} = \begin{pmatrix} \lambda_1 & 0 \\ 0 & \lambda_2 \end{pmatrix}\begin{pmatrix} x \\ y \end{pmatrix}. \qquad (1.22)$$

It is easy to see that $x = 0$ and $y = 0$ are invariant curves that extend beyond the invariant fixed point at $(x, y) = (0, 0)$. This is because the matrix in (1.22) has the unit vectors along the x- and y-axes as eigenvectors. Therefore the velocity vectors of points located along these two lines are tangent with these lines: the states that are located in these curves stay on them. This is illustrated in Fig. 1.12a.

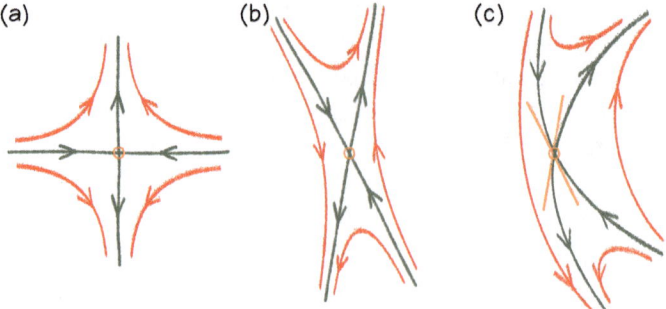

Figure 1.12: (a) Phase portrait of system (1.22). The green solid lines are the invariant manifolds, red lines are examples of trajectories. (b) Phase portrait for a nondiagonal linear system. Invariant manifolds are straight lines aligned on the eigendirections of the linear operator. (c) Phase portrait for a nonlinear system, with the nonlinear terms creating curvature in the invariant manifolds. The orange lines are the eigendirections of the linearization of the system in the vicinity of the fixed point.

For a general (i. e., nondiagonal) system $\dot{\mathbf{X}} = \mathbf{MX}$, the straight lines aligned along the eigendirections of the matrix defining the system are also invariant sets, from the very definition of an eigenvector (Fig. 1.12b). As invariant sets, they form barriers that the nearby trajectories cannot cross.

Let us now turn to a nonlinear system $\dot{\mathbf{X}} = \mathbf{MX} + N(\mathbf{X})$ whose linear part coincides with the system in Fig. 1.12b (i. e., $N(0) = 0$, $(\partial N/\partial \mathbf{X})(0) = 0$). If we consider a continuous deformation $\dot{\mathbf{X}} = \mathbf{MX} + \xi N(\mathbf{X})$, then we see that as ξ is increased from 0 to 1, the invariant straight lines in Fig. 1.12b will deform into invariant curves that are tangent to the eigendirections at the origin (where the nonlinear part is negligible), as shown in Fig. 1.12c. We will return to this discussion in Section 2.3, where we will study the invariant manifolds of fixed points.

Thus it appears that the influence of a fixed point is felt in a large part of the phase space due to the invariant curves that pass through it and serve as separatrices.

When there is a pair of complex conjugate eigenvalues, there is an invariant two-dimensional surface tangent to the corresponding eigenplane at the fixed point (Chapter 2).

More generally, when working in higher dimension, it is useful to group the eigenspaces of the linearized flow around the fixed points and accordingly the associated invariant curves and planes according to their stability properties: do trajectories contained in them converge to the fixed point or diverge from it? We will see in Chapter 2 that this is related to the sign of the real part of the associated eigenvalue.

More precisely, the stable ($W^s(\mathbf{X}^*)$) and unstable ($W^u(\mathbf{X}^*)$) manifolds of a fixed point $\mathbf{X}^*$ gather states whose trajectories converge to $\mathbf{X}^*$ in the future or in the past, respectively:

$$W^s(\mathbf{X}^*) = \left\{ \mathbf{X} \in \mathcal{S} : \lim_{\tau \to \infty} \phi_\tau(\mathbf{X}) = \mathbf{X}^* \right\}, \tag{1.23}$$

$$W^u(\mathbf{X}^*) = \left\{ \mathbf{X} \in \mathcal{S} : \lim_{\tau \to -\infty} \phi_\tau(\mathbf{X}) = \mathbf{X}^* \right\}. \tag{1.24}$$

As with the invariant curves and surfaces mentioned earlier, the stable and unstable manifolds are invariant by construction, as the trajectory of any point inside them remains inside. When a two-dimensional fixed point is partially stable and partially unstable, these manifolds are one-dimensional and therefore divide the phase plane. States that are on one side of an invariant manifold stay on the same side.

In some cases, there is one or several eigenvalues with a zero real part, which is a peculiar situation that generally disappears through a small parameter change. Trajectories in the corresponding eigenspaces close to the fixed point neither converge nor diverge and therefore evolve on infinitely long times, as do the trajectories in the associated invariant surfaces. The dynamics of the system is then driven by nonlinear terms and is highly nontrivial. Typically, there occurs a qualitative change associated with a change in the number and/or stability of invariant sets, which is called a *bifurcation*. Bifurcations will be studied in Chapter 3, after we learn to study stability in Chapter 2.

Application to the pendulum examples – In the case of the damped pendulum, all the trajectories are attracted by the fixed point $(0, 0)$. Consequently, in this example, there is only one stable manifold, the whole phase space.

As discussed previously, the conservative pendulum is a particular example as it does not have attractors. Still, it presents fixed points with in particular the point $(\pi, 0)$ (which is identical to $(-\pi, 0)$ because of the 2π-periodicity of θ). The separatrices (Section 1.2.2) are the invariant manifolds associated with this fixed point (see Fig. 1.7). A distinctive feature of those manifolds is that they connect the fixed point to itself. The trajectories that start from one fixed point and return to the same fixed point (resp., another fixed point) are called *homoclinic* (resp., *heteroclinic*) orbits.

1.3.2.c Limit cycles

There are other important one-dimensional invariant sets that exist in the form of closed trajectories, namely *limit cycles*.

In dimension one, we saw that the velocity $\dot{x}$ of the state variable can never change sign, only converge to zero. When the dimension of phase space is two or higher, some trajectories can return exactly to their initial condition in a finite time after visiting other places in the phase space. When restricted to these closed orbits, the motion is periodic, as it repeats infinitely. In fact, it can be shown that when the dynamics takes place in a bounded region of the phase space, the existence of closed orbits is as natural as the existence of fixed points when certain conditions are met, as discussed in Section 1.2.4. This can be shown using arguments similar to those we used for fixed points and which depend on very mild assumptions, as we will see in Section 4.1.2.b.

As with other one-dimensional invariant sets, a closed orbit organizes the flow around it, as it forms a barrier that cannot be crossed by other trajectories in the two-dimensional phase space. As stated by the Jordan curve theorem, it divides the plane into an "interior" region and an "exterior" region. The trajectories that are inside (resp., outside) the cycle remain inside (resp., outside). Note that the nonintersection theorem also applies to the closed orbit, which cannot cross itself or another limit cycle.

Together, fixed points dressed with their invariant manifolds and closed orbits organize the phase plane because they are barriers that cannot be crossed due to the nonintersection theorem, as illustrated in Fig. 1.13.

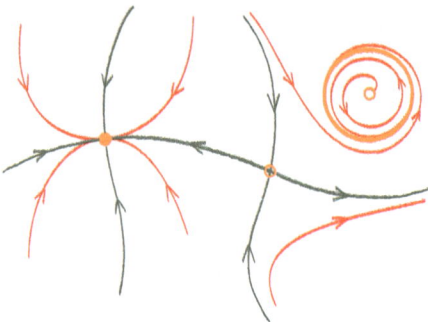

Figure 1.13: Two-dimensional phase portrait showing three fixed points: a stable node (left) as well as a saddle point (center) and a focus point (right), both unstable. The invariant manifolds of the first two points (green lines) and a limit cycle encircling the focus point (orange line) are invariant structures that guide trajectories (red) and partition the phase plane.

Application to the pendulum – In the case of the conservative pendulum, the closed orbits are not attractors and thus cannot illustrate the concepts we just introduced. The damped pendulum, on the other hand, only has the fixed point $(0,0)$ for attractor. In fact, if no energy is injected into a dissipative system to compensate for the loss of energy by damping, then no complex dynamics will emerge. The system will just converge to a static position corresponding to a minimization of energy.

Consequently, to observe a limit cycle in our red thread example, we have to consider the driven damped pendulum described by Eqs. (1.10). As discussed in Section 1.1.3, this system is in fact three-dimensional in its autonomous form. Limit cycles living in the plane $(\theta, \dot{\theta})$ can be observed, as shown in Fig. 1.14a, but much more complicated behaviors exist depending on the value of the parameters. For example, Fig. 1.14b shows also a limit cycle but a rather special one. First, the trajectory seems to cross itself, but this is only due to the projection of the 3D dynamics into a plane. It is actually surprising that the oscillation period is not equal to the driving frequency as in Fig. 1.14a but to the half of this frequency. Interestingly, this is a form of symmetry breaking experienced by the spontaneous magnetization of a magnetic material in the paramagnetic-ferromagnetic phase transition. Indeed, the system is invariant under time translation by multiples of $T = \frac{2\pi}{\omega}$, which corresponds to a certain symmetry group, but the solution is only invariant under time translation by multiples of $2T$, which is a subgroup of this group.

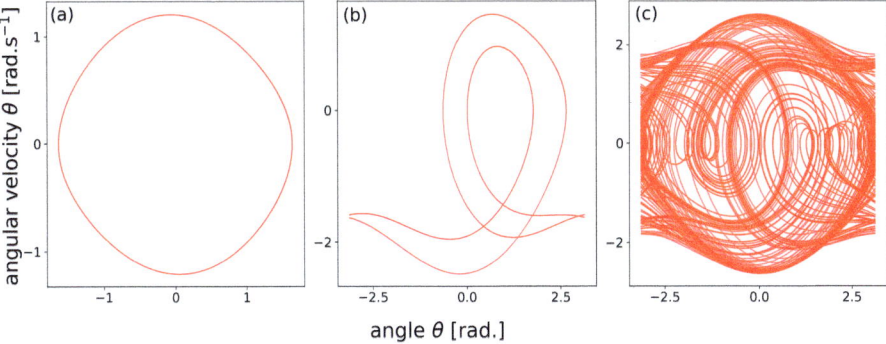

Figure 1.14: Trajectories projected in the plane $(\theta, \dot{\theta})$ of the dynamical system (1.10) with $g/l = 1.0$, $\gamma = 0.5$, $\omega = 0.667$; (a) $M = 0.7$, (b) $M = 1.45$, (c) $M = 1.5$.

1.3.3 Dynamics in dimension three and higher

In Chapter 5, we will study two very different types of complex behavior, which can only be observed in dimension three and higher, quasi-periodicity and chaos.

Since one elementary oscillation requires at least two dimensions, the interaction/combination of two or several oscillating degrees of freedom can only unfold in dimension three or higher. As we will see in Section 4.3.2 and Chapter 5, the dynamical nature of a multioscillatory regime depends on the ratios between the different frequencies involved. If these ratios are all rational numbers, then there is a global period for the dynamics: we have a periodic orbit, possibly with a complex geometry that only fits in a high-dimensional phase space, as with the limit cycle showed in Fig. 1.14b. A *quasi-periodic* regime is also possible when the frequency ratios are irrational. The system then evolves on a new type of invariant object of intrinsic dimension 2 or greater, an invariant torus (Fig. 1.15a).

Evolving in a 3D phase space allows still another type of recurrent behavior besides stationary, periodic, or quasi-periodic behavior. As with quasi-periodic regimes, *deterministic chaos* is an irregular behavior that returns infinitely many times and arbitrarily close to previously visited states without closing the trajectory exactly. However, its signature is a permanent instability manifesting itself as *sensitivity to initial conditions*. A chaotic system explores still another type of invariant object with complex structure, a *strange attractor* (Fig. 1.15b), which we will also study in Chapter 5.

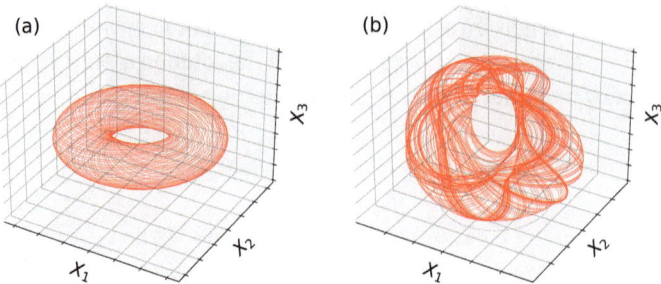

Figure 1.15: (a) An example of a quasi-periodic trajectory evolving on a torus in 3D. (b) Representation in 3D of the solution of Eqs. (1.10) shown in Fig. 1.14c. The following phase-space coordinates have been used: $X_1 = \rho \cos \omega t$, $X_2 = \rho \sin \omega t$, and $X_3 = \dot{\theta}$ with $\rho = 1.7 + \sin \theta$.

Application to the pendulum – Let us now consider the driven pendulum of Eq. (1.10), which is a 3D system. We have previously discussed the existence of limit cycles in this system (Fig. 1.14a,b), but depending on the parameter values, its dynamics can be much more complex.

Figure 1.14c shows an orbit obtained by numerical integrations of system (1.10) projected in the $(\theta, \dot{\theta})$ plane. The behavior of the driven pendulum is very surprising because the system does not settle on a closed trajectory. This is not a transient but a truly asymptotic regime, obtained after discarding the beginning of the numerical simulation. The same observation would hold if we built an actual experiment. Moreover, the motion in the phase space is not random. A structure clearly emerges when the trajectory is plotted for a sufficiently long time, but it is very complex to decipher. The representation in the 3D phase space Fig. 1.15b (using convenient variables as specified in the caption) reveals a global structure. This type of behavior is in fact termed "deterministic chaos,"[2] and these "weird" structures in the phase space are examples of strange attractors.[3]

It is quite mind-puzzling that a system as "simple" as a damped driven pendulum can exhibit such a complex behavior. This system is usually studied in undergraduate classes without any mention of such strange behavior. This is because only the linearized version of the system is generally studied, when the oscillations are small. Linear systems appeal to the physicist because they obey a superposition principle, which means that any solution can be expressed as a linear combination of eigenstates of some operator. This is a key to central equations of physics, such as the Maxwell or Schrödinger

---

2 James Yorke, one of the most prolific chaos scientists, is usually credited with using the word "chaos" for the first time in a seminal paper titled "Period-3 implies chaos" (Li and Yorke, 1975), although the term chaotic had already appeared in a paper by Ruelle and Takens (Ruelle and Takens, 1971). James Yorke has coauthored important advances in chaos theory, building a bridge between mathematics and physics. He was awarded the Wolf Prize in Physics for his work.

3 The term "strange attractor" was coined by David Ruelle (Ruelle, 1980), a mathematical physicist who played a key role in identifying central concepts of chaotic dynamics and its links with statistical physics.

equations. Thus linear systems are amenable to analytical approaches. However, a linear system will never exhibit chaotic behavior or even oscillations at a different frequency than the imposed one. Actually, it can be shown that any linear system can be reduced to a combination of two-dimensional oscillations.

A key ingredient to observe "exotic" behavior such as displayed in Fig. 1.14c and also in Fig. 1.14b is nonlinearity, which in our example is given by the terms "$\sin X_1$" and "$\sin X_3$." Another necessary ingredient is the phase space of dimension 3 or higher. A nonlinear system confined in a plane cannot exhibit a complex behavior, even as simple as the periodic motion seen in Fig. 1.14b because of the constraints imposed by the nonintersection theorem (Section 1.3.2). Yet, nonlinearity is not a sufficient condition to observe chaos: there are nonlinear differential equations that never exhibit complex behavior regardless of the parameter values.

As in many scientific problems, complexity is not an obstacle to understanding if we identify the questions that are actually important to answer and those that are irrelevant. Nonlinear systems can display irregular behavior, but it is still possible to analyze and classify it, unveiling universal properties of nonlinear behavior by looking at it from the right angle. To achieve this, we need to go beyond mere numerical simulation of a differential equation system and to gather concepts that will allow us to think about nonlinear dynamical systems, as we will see in the following chapters.

1.4 Dimensionality reduction and discrete-time dynamical systems

Until now, we have considered systems that evolved continuously in time and thus were described by systems of ordinary differential equations. However, it is generally not useful to follow the system trajectory down to the smallest detail. We may prefer to restrict ourselves to sampling the state occasionally, while still capturing relevant information about the dynamics at play. We already encountered this concept when we introduced the notion of a flow ϕ_τ that transforms a state $\mathbf{X}(t)$ into a state $\mathbf{X}(t + \tau)$ on a time interval τ (Section 1.2.2), essentially reformulating a differential system into a mapping from the phase space into itself.

However, this concept is especially fruitful when it leads to a reduction of the dimensionality of phase space. After all, if we are interested only in the nature of the asymptotic behavior of a system, then we do not need to know how all the regions of the phase space are visited. It then makes sense to focus on when the system visits a subset of the phase space involved in the asymptotic dynamics and on what are the states successively visited in this subspace. If this subspace has a lower dimension than the original phase space, then these states will form a discrete sequence $\{\phi(t_k)\}$, where t_k correspond to the times of crossings.

1.4.1 Poincaré section

This idea is formalized in the concept of a *Poincaré section*, named after the mathematician that invented many concepts in dynamical systems and topology. If the phase space is d-dimensional, then we consider a surface Σ of dimension $d-1$ that is relevant to the asymptotic dynamics (any trajectory starting from the surface will cross it again infinitely many times). We also require that Σ is everywhere transverse to the velocity vector $\mathbf{F}(\mathbf{X})$. Fixing a normal $\mathbf{n}_\Sigma(\mathbf{X})$ to the surface, we moreover restrict ourselves to the part Σ_+ of the surface where $\mathbf{n}_\Sigma(\mathbf{X}).\mathbf{F}(\mathbf{X}) > 0$: the system crosses Σ_+ always coming from the same side. It can be shown that if $\mathbf{X}$ is a recurrent point that is not a fixed point, then such a surface can always be defined in a neighborhood of $\mathbf{X}$. In many cases, it is possible to extend this surface in the entire phase space.

Now, starting from one point $\mathbf{X}_0 = \mathbf{X}(t_0) \in \Sigma_+$, since Σ_+ is of dimension $(d-1)$, the continuous trajectory $\mathbf{X}(t)$ starting from $\mathbf{X}_0$ will intersect Σ_+ at times t_k in a discrete and infinite sequence of states $\mathbf{X}_k = \mathbf{X}(t_k)$ (Fig. 1.16). The time intervals between two intersections will typically depend on the position in the section plane; yet what is relevant is only how an intersection is mapped to the next intersection. Still, useful information can sometimes be obtained by studying how the time-of-flight between two crossings depends on the previous time-of-flight.

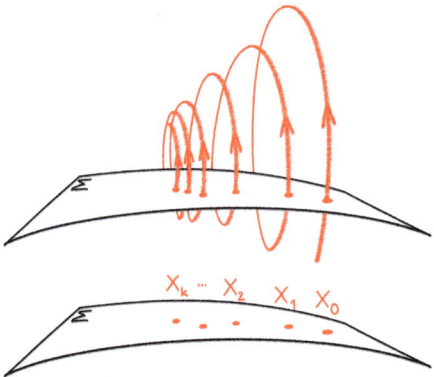

Figure 1.16: Concept of a Poincaré section. Given a suitably chosen surface Σ, the trajectory of the system in the phase space (in red) will intersect it an infinite number of times. Keeping only the intersections where the surface is crossed from a given side to the other, we obtain a sequence of points $\{X_0, X_1, \ldots, X_k, \ldots\}$, which form a Poincaré section.

This operation of reducing the continuous trajectory $\{\mathbf{X}(t) : t \in \mathbb{R}\}$ to a discrete sequence $\{\mathbf{X}(t_k)\}$ in a surface of lower dimension is called a *Poincaré section*.

Because the system is deterministic, there exists a mapping $\mathcal{P}$ that relates the successive intersections $\mathbf{X}_k$ of the trajectory $\mathbf{X}(t)$ with Σ_+ through $\mathbf{X}_{k+1} = \mathcal{P}(\mathbf{X}_k)$. This *Poincaré map* $\mathcal{P}$ (also called the first return map) can be defined as sending any point $\mathbf{X} \in \Sigma_+$ to the first return of its trajectory in Σ_+:

$$\forall \mathbf{X} \in \Sigma_+, \quad \mathcal{P}(\mathbf{X}) = \phi_\tau(\mathbf{X}) \quad \text{such that } \tau = \min\{t > 0 : \phi_t(\mathbf{X}) \in \Sigma_+\}$$

A Poincaré section is a powerful concept because it reduces the dimensionality of the system under study. In studying 3D complex flows, we will be able to restrict ourselves to consider how a surface is mapped to itself, replacing the integration of a system of ODEs by the application of a mapping.

Poincaré sections are especially useful in the neighborhood of a recurrent point, where the Poincaré map associated with a section that is transverse to the flow will typically have a fixed point near the recurrent point. In particular, we will harness Poincaré sections to study the stability and bifurcations of periodic orbits in Section 4.2.2. The concept of a Poincaré section is also at the heart of the proof of the Poincaré–Bendixson theorem that we will study in Section 4.1.2.b.

A Poincaré section is particularly easy to define and useful in the case of systems forced by an external periodic signal of period T_0. The phase φ of the forcing signal is then a natural variable since the governing equations typically depend on it. By carrying out a stroboscopic sampling at times $t = kT_0 + t_0$, we essentially build a Poincaré section of constant phase φ.

Application to the pendulum – The concept of Poincaré section can be applied first to the damped pendulum. In that case, as the phase space is of dimension 2, the surface of section is a 1D set. A Poincaré section of the trajectory can be obtained using the successive intersections of the orbit with the x-axis when $\ddot{\theta} > 0$. The series of points θ_k obtained converges exponentially to 0 (Fig. 1.17).

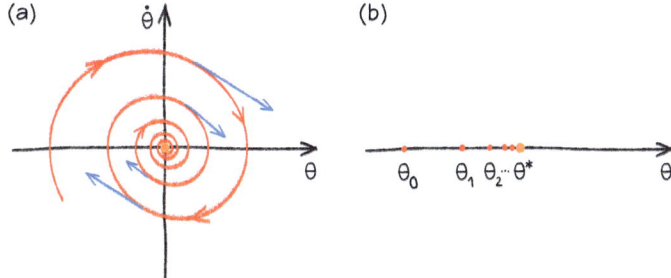

Figure 1.17: (a) An example of trajectory of a damped pendulum in phase space. (b) Poincaré section of the trajectory.

Let us now consider the driven pendulum of Eqs. (1.10). In this 3D system, the third coordinate, X_3 (= ωt), is a phase and is thus 2π-periodic. As discussed above, it is common for such a system to measure the values of X_1 (= θ) and X_2 (= $\dot{\theta}$) at discrete time steps fixed by the forcing period, as in a stroboscopic measurement. We will see examples of this type in Chapter 4. The resulting dynamics is discrete and is shown in Fig. 1.18

with the corresponding asymptotic dynamics in the phase space in inserts. Note that in opposition with the previous example of the damped pendulum, only the asymptotic dynamics are plotted in Fig. 1.18.

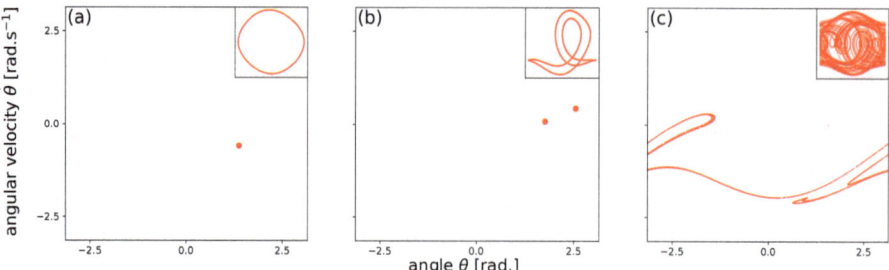

Figure 1.18: Poincaré sections of the dynamical system (1.10) with $g/l = 1., \gamma = 0.5, \omega = 0.667$ and (a) $M = 0.7$, (b) $M = 1.45$, (c) $M = 1.5$. The corresponding trajectory in phase space for each value of M and corresponding to Fig. 1.14 are recalled in the inserts.

In the case of the limit cycle of Fig. 1.18a, for which the pendulum oscillates at the imposed frequency ω, a single point corresponding to this period-1 orbit is obtained in the Poincaré section. In the case of the limit cycle of Fig. 1.18b, which corresponds to an oscillation at half the driving frequency, two points are obtained in the Poincaré section corresponding to the two loops of the orbit. Finally, in the case of the chaotic behavior of Fig. 1.18c, the Poincaré section shows a linear structure, which seems to be folded. The stretching and folding mechanisms that underlie the structure of strange attractors will be discussed in Chapter 5.

1.4.2 Recurrence systems

Since Poincaré sections allow us to reformulate the study of differential systems in terms of mappings, it is useful and relevant to study mappings for themselves, reducing the mathematical complexity while preserving the dynamical complexity. Such mappings typically have the form

$$\mathbf{X}_{n+1} = \mathbf{F}(\mathbf{X}_n). \tag{1.25}$$

A fixed point of Eq. (1.25), satisfying $\mathbf{X}^* = \mathbf{F}(\mathbf{X}^*)$, is generally termed a period-1 orbit, given that such a mapping is the analog of a Poincaré map. More generally, a periodic orbit of period p satisfies $\mathbf{X}_{n+p} = \mathbf{F}^p(\mathbf{X}_n) = \mathbf{X}_n$.

Considering a one-dimensional recurrence $x_{n+1} = f(x_n)$, the stability of a fixed point $x^* = f(x^*)$ can be determined following a similar approach as in Section 1.3.1 by studying the time evolution of a deviation δx from the fixed point x^*. Indeed, we have

$$x^* + \delta x_{n+1} = f(x^* + \delta x_n) \simeq f(x^*) + f'(x^*)\delta x_n = x^* + f'(x^*)\delta x_n$$

and, by recurrence, $\delta x_n = (f'(x^*))^n \delta x_0$. Consequently, perturbations converge to zero when $|f'(x^*)| < 1$, which is thus the condition for the fixed point to be stable. The number $\mu = f'(x^*)$ is called the *multiplier* of the fixed point.

In higher dimensions, the time evolution of a perturbation is governed by

$$\delta \mathbf{X}_{n+1} = \left(\frac{\partial \mathbf{F}}{\partial \mathbf{X}}\right)_{\mathbf{X}=\mathbf{X}^*} \delta \mathbf{X}_n,$$

where the eigenvalues of the Jacobian matrix $\partial \mathbf{F}/\partial \mathbf{X}$ should all have moduli smaller than one for the fixed point to be stable.

As we will see in Chapter 6, the one-dimensional logistic map

$$x_{n+1} = r x_n (1 - x_n), \quad x \in \mathbb{R}, \tag{1.26}$$

can display extremely complicated dynamics. When r is sufficiently small, a unique stable fixed point is observed. For some value of r, this fixed point becomes unstable giving birth to a period-2 orbit, which experiences the same scenario at higher values of r. Following a cascade of such events, we then observe irregular behavior known as deterministic chaos. Figure 1.19 shows a sample orbit of (1.26).

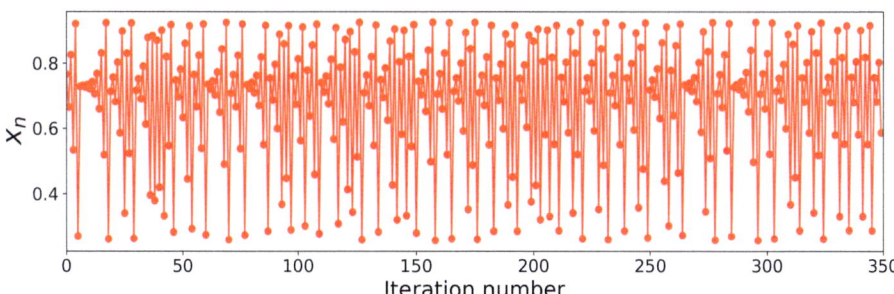

Figure 1.19: Successive iterates obtained with the logistic map (1.26) starting from $x_0 = 0$ for $r = 3.7$. This behavior is an example of deterministic chaos.

In the case of 1D maps, the successive values taken by an iterated mapping such as $x_{n+1} = f(x_n)$ can be represented graphically using the graph of the function $f(x)$ and the diagonal of this graph, as illustrated in Fig. 1.20. Starting from an initial condition x_0, we draw a line to the graph of $f(x)$ to obtain $f(x_0) = x_1$ along the vertical axis. Since this point is used for the next iterate, we draw a line to the diagonal $y = x$ to obtain its location along the horizontal axis. We then repeat the construction infinitely, drawing alternatively a line to the graph and then to the diagonal.

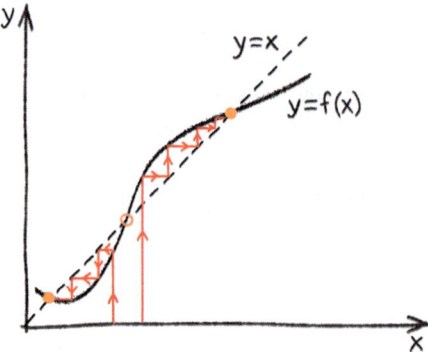

Figure 1.20: Graphical method of construction of iterates $x_{n+1} = f(x_n)$ using the graph of the function $f(x)$ and the line $y = x$.

The intersection points of the curve $y = f(x)$ and of the line $y = x$ give the fixed points of the system as they obey to $x^* = f(x^*)$. The stability of the fixed point can be deduced graphically by checking the slope of the tangent to the curve in x^*.

Figure 1.21 shows the different situations that can arise depending on how the tangent to the graph of $f(x)$ is placed compared to the straight lines $y = x$ (constant slope of 1) or $y = -x$ (constant slope of –1). This illustrates that a fixed point is indeed stable when $|f'(x^*)| < 1$ and unstable otherwise.

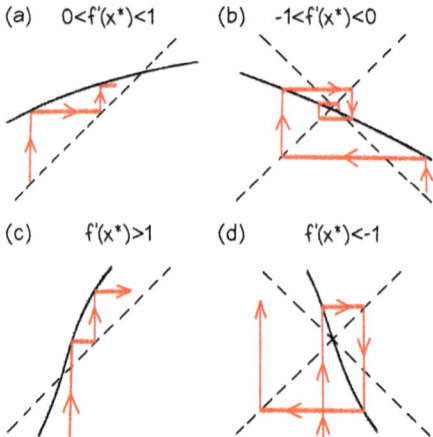

Figure 1.21: Illustration of the construction of iterates for the different values of the slope of the tangent at the fixed point $f'(x^*)$.

1.5 Conclusions

In this introductory chapter, we have recalled that there are many interesting systems around us that evolve deterministically in time, and we have begun to build a general theory to describe their behavior. Importantly, a geometrical description of the dynamics by representing the system in a state space allowed us to obtain a global view of the behavior of the system. We encountered important concepts such as the velocity vector in the phase space, the flow transforming regions of phase space into other regions, invariant sets organizing the phase space (fixed points, limit cycles, strange attractors, ...), Poincaré sections, ...

We have seen that for dissipative systems, all the trajectories converge toward attractors that carry the asymptotic dynamics. An attractor is a subset of the phase space that is both invariant and stable (attracting). However, these properties may be modified when a control parameter of the system varies. In the next chapters, we will study how invariant sets can change their stability, appear or disappear.

In Chapter 2, we will learn how to carry a stability analysis of an invariant set. This will mostly be done for fixed points, but at the end of the chapter, we will give indications on how to extend the concepts to time-varying trajectories.

In Chapter 3, we will study how the existence or stability of fixed points can be modified when a control parameter is modified. The analysis will be done for one-dimensional dynamical systems; however, we will show that it remains relevant for higher-dimensional systems. The case where a fixed point loses stability by giving birth to a periodic orbit will be considered in Chapter 4.

Exercises

Trajectories in the phase space

System 1
Consider the following system:

$$\begin{cases} \dot{x} = ax, \\ \dot{y} = -ay. \end{cases}$$

1. Plot the vector field in the phase space for $a > 0$.
2. Integrate the dynamics for the initial condition $(x(0), y(0)) = (x_0, y_0)$ and show that the Cartesian equation of the trajectories is $y(x) = \frac{x_0 y_0}{x}$. Draw several trajectories.
3. Does the flow preserve area (i. e., preserve 2D-volume)?

System 2
Same questions for the system

$$\begin{cases} \dot{x} = ax, \\ \dot{y} = ay, \end{cases}$$

with the Cartesian equation $y(x) = \frac{y_0}{x_0}x$ for the trajectories.

System 3
Same questions for the system

$$\begin{cases} \dot{x} = -ay, \\ \dot{y} = ax, \end{cases}$$

with the Cartesian equation $x^2 + y^2 = x_0^2 + y_0^2$ for the trajectories.

System 4
Consider the system

$$\begin{cases} \dot{x} = ay, \\ \dot{y} = ax. \end{cases}$$

Show that the change of coordinates

$$\begin{cases} \dot{x} = x + y, \\ \dot{y} = x - y, \end{cases}$$

allows us to recover system 1 and deduce the Cartesian equation for the trajectories.

One-dimensional dynamics

A mathematical example
The purpose of this exercise is to show the power of the graphical tools presented in this chapter in contrast to the thorough calculations of all possible cases.
Consider the following system:

$$\frac{dx}{dt} = 1 - x^2 \quad \text{with } x(0) = x_0.$$

1. Show that the solution of the differential equation is

$$x(t) = \frac{\frac{x_0+1}{x_0-1}e^{2t} + 1}{\frac{x_0+1}{x_0-1}e^{2t} - 1}.$$

2. If $x_0 \in]-1, 1[$, then can the denominator $\frac{x_0+1}{x_0-1}e^{2t} - 1$ cancel? What is the asymptotic behavior of $x(t)$ as $t \to +\infty$ for this set of initial conditions?
3. The same question for $x_0 \in]1, \infty[$.
4. The same question for $x_0 \in]-\infty, -1[$.
5. What is the dynamics in the particular cases $x_0 = 1$ and $x_0 = -1$.

Now we want to recover those results using only graphical methods.
6. Draw the function $f(x) = 1 - x^2$.
7. Using Section 1.3.1, find the fixed points of the system and the possible behaviors as a function of the initial conditions.

Verhulst model of population dynamics

The Belgian mathematician Pierre-François Verhulst has introduced at the begin-ning of the nineteenth century the following dynamical equation to model population growth:

$$\frac{dN}{dt} = rN\left(1 - \frac{N}{N^*}\right),$$

where $N(t)$ is the number of individuals at time t, $r > 0$ is the maximum growth rate, and N^* is the carrying capacity, i. e., the maximum number of individuals that the environment can support.
1. Draw the function $f(N) = rN(1 - \frac{N}{N^*})$.
2. Deduce from the graph the fixed points of the system and their stability.
3. Draw the evolution of the population with time $N(t)$ for several initial conditions.

Two-dimensional dynamics

A mathematical example

Consider the following bidimensional system:

$$\begin{cases} \dot{x} = y - 2x, \\ \dot{y} = x + x^2 + y. \end{cases}$$

1. What is the Jacobian matrix $(\frac{\partial F(X)}{\partial X})_{ij} = \frac{\partial F_i(X)}{\partial X_j}$ of this system?
2. Is the system conservative or dissipative?
3. Draw the nullclines in the phase space and the vector field along those lines.
4. Draw general direction of the vector field in the different areas of the phase space delimited by the nullclines.
5. What are the fixed points of the system? Can you comment on their stability?

Three-dimensional dynamics

Topological study of the Lorenz model

Edward N. Lorenz (1917–2008) was a meteorologist who evidenced sensitivity to initial conditions in chaotic systems, a concept we will study in Chapter 5. He worked on a very simplified model of convection which now is named after him (Lorenz, 1963):

$$\dot{X} = \text{Pr}(Y - X),$$
$$\dot{Y} = -XZ + rX - Y,$$
$$\dot{Z} = XY - bZ.$$

The variable X is related to the velocity field of the fluid, whereas the variables Y and Z are linked to the temperature field, Pr is the Prandtl number (a dimensionless number defined as the ratio of the kinematic viscosity to the thermal diffusivity), r is proportional to the Rayleigh number (another dimensionless number characterizing heat transfer in a fluid), and b is a geometrical factor linked to the thickness of the convection layer. The three parameters are strictly positive. We denote the velocity vector field by $\mathbf{F} = (\dot{X}, \dot{Y}, \dot{Z})$ with the time derivatives specified by the equations above.

1. Compute the divergence of $\mathbf{F}$. Conclude on the nature of the system (dissipative vs. conservative).

We consider in the phase space the ellipsoid defined by

$$g(X,Y,Z) = \frac{X^2}{2\,\text{Pr}} + \frac{Y^2}{2} + \frac{Z^2}{2} - (r+1)Z - \mu = 0,$$

where μ is a constant.

2. Explain with the help of a schematic the meaning of the quantity

$$P = \dot{X}\frac{\partial g}{\partial X} + \dot{Y}\frac{\partial g}{\partial Y} + \dot{Z}\frac{\partial g}{\partial Z}.$$

3. Compute $P(X,Y,Z)$. What is its sign for large values of X, Y, Z? Conclude that it is always possible to find an ellipsoid defined by a constant μ large enough for P to be negative at all the points of the surface of the ellipsoid.
4. Conclude on the possible asymptotic trajectories of the Lorenz system.

Recurrence maps

As discussed in Section 1.4.2, complex discrete dynamics can be observed by iterating one-dimensional maps. In this exercise, we will study some aspects of the dynamics of the logistic map that Mitchell J. Feigenbaum (1944–2019) used to demonstrate universal features of one-dimensional maps (Feigenbaum, 1980):

$$x_{n+1} = f(x_n) = rx_n(1 - x_n),$$

where $r > 0$.

1. Draw the graph of the function $f(x)$ for $r > 0$. Determine the position of the maximum x_m and the value $f(x_m)$.

2. For the dynamics to be bounded, we must restrict the study of the map to an interval $[a, b]$ such as $[f(a), f(b)] \subset [a, b]$. Show that such a property is verified for $x \in [0, 1]$ if the variations of r are restricted to an interval whose limits will be given.

3. Determine analytically the fixed point(s) of f, their domain of existence, and their stability. The nontrivial fixed point will be further denoted x^*.

4. Draw the graph of f and the bisectrix $y = x$ for $r \leq 1, r = 1$, and $r \geq 1$. Find graphically the results of the previous question.

5. Determine the value r_1 for which x^* loses its stability.

6. Draw the graph of f as well as the bisectrix $y = x$ for $r \geq r_1$ and construct graphically the iterated points starting from any initial condition. Comment on the dynamics observed.

7. Write a Python script to plot the iterates of $f(x)$ and recover the results of the previous questions.

Numerical exercises

The objective of these exercises is to write a Python script that will evolve throughout the reading of this book. The elementary functions introduced in this first part will be necessary to perform more elaborate numerical studies later on. The prerequisite for the part of the exercises devoted to numerical studies is to have a basic knowledge of algorithms in general and of the Python language in particular.

Numerical integration of dynamical systems

To integrate a dynamical system and then plot it in the state space, we are going to use the function `solve_ivp` of the package `scipy.integrate`.

Our first goal is to write a script to plot Fig. 1.18.

1. Create a script in Python, import `solve_ivp` and pyplot.

2. Define a function `forced_pendulum(t,x)` using Eqs. (1.10).

3. Define the values of the parameters and the initial condition.

4. Integrate the system using `solve_ivp`. In this book, we do not treat the different integration schemes that can be used to integrate an ODE system depending on the stiffness of the problem. For this example, we can use an explicit Runge–Kutta method such as RK45, but for stiffer problems, you may want to use other methods (see the detailed description of the `solve_ivp` function).

5. Plot x[1] as a function of x[0] for different values of the parameters.
6. Plot the Poincaré section of the dynamics by using the value taken every $2\pi n/\omega$.

Your script should look something like:

```python
# Importing Packages
import numpy as np
import matplotlib.pyplot as plt
from scipy.integrate import solve_ivp


# Dynamical system
def F_pendulum(t,x):
    x0_dot = x[1]
    x1_dot = -g_over_l*np.sin(x[0]) - gamma*x[1] + M*np.sin(x[2])
    x3_dot = omega
    return [x0_dot, x1_dot, x3_dot]


# Parameters
g_over_l = 1.
gamma = 0.5
M = 0.7
omega = 0.667


# Initial Conditions
X0 = [np.pi/2,0,0]


# Values of time where the function will be evaluated after transient
has elapsed
t_a = np.arange(4500,5000,1e-2)
# integration of the system
sol = solve_ivp(F_pendulum,[0, 5000], X0,method = 'RK45',t_eval = t_a,
first_step=1e-4)
theta = sol.y[0,:]
# taking into account pi periodicity of the angle variable
theta = (theta + np.pi)%(2*np.pi) - np.pi
thetadot = sol.y[1,:]


# Values of time where the function will be evaluated to perform
a Poincaré section based on a stroboscopic methd
t_p = np.arange(4000,8000,2*np.pi/omega)
# Solving ODE transitory
sol = solve_ivp(F_pendulum,[0, 8000], y0,method = 'RK45',t_eval = t_p,
first_step=1e-4)
```

```
theta_p = sol.y[0,:]
theta_p = (theta_p + np.pi)%(2*np.pi) - np.pi
thetadot_p = sol.y[1,:]

# Plotting the attractor and the Poincaré section side by side
fig, ax = plt.subplots(nrows=1, ncols=2, sharex=True, sharey=True,
figsize=(16, 8))
ax[0].plot(theta,thetadot,'.')
ax[1].plot(theta_p,thetadot_p,'.')
```

Vector fields

Another tool of interest is the representation of a vector field in the plane. The Python function allowing to draw a field of arrows is quiver from the matplotlib.pyplot package. The following script performs this task.

```
# Importing Packages
import numpy as np
import matplotlib.pyplot as plt

# Vector field
def F_pendulum(t,x):
    x0_dot = x[1]
    x1_dot = -g_over_l*np.sin(x[0]) - gamma*x[1]
    return [x0_dot, x1_dot]

# Parameters
g_over_l = 1.
gamma = 0.5

# Definition of the mesh on the nodes of which the vector field will be
computed
x = np.arange(-np.pi, np.pi, 0.3)
y = np.arange(-4, 4, 0.5)
X, Y = np.meshgrid(x, y)

V = F_pendulum((X,Y))

# Plot
fig, ax = plt.subplots(figsize =(14, 8))
ax.quiver(X, Y, V[0],V[1])
```

Exercise:
1. On the same plot, superimpose the vector field, the nullclines, and several trajectories of the conservative swinging pendulum.
2. The same question for the damped pendulum.

Rössler attractor

In Chapter 5, we will use a system proposed by Otto Rössler (1976) as a benchmark to study strange attractors. O. Rössler developed his model to obtain the simplest set of equations exhibiting chaotic behavior with a clear intuition of the geometric mechanisms generating chaos in phase space. The Rössler system reads

$$\dot{x} = -y - z, \tag{1.27}$$
$$\dot{y} = x + ay,$$
$$\dot{z} = b + z(x - c).$$

1. Integrate this system of equation for $a = b = 2.0$ and $c = 5.0$ to obtain the Rössler attractor shown in Fig. 1.15b.
2. Integrate the system for $c = 2.5$, $c = 3.5$, and $c = 4$ while keeping $a = b = 0.2$. Plot the resulting asymptotic dynamics in the (x, y) plane.

2 Stability analysis

In this chapter, we study how to perform a linear stability analysis of an ordinary differential equation system in the neighborhood of a fixed point to determine whether close trajectories converge to the fixed point or diverge from it.

This method proceeds by linearizing the nonlinear system around the fixed point, exactly as the graph of a function can be approximated by its tangent in the vicinity of a point. We already followed this strategy in Section 1.3.1 to determine the stability of the fixed point of a 1D system and gave a hint of how it should be done in a 2D system in Section 1.3.2. Thus we will first take up the 2D calculation where we had left it in Section 1.3.2 and discuss all possible qualitative behaviors of trajectories around fixed points in a plane. We will then move to a phase space of arbitrary dimension, and we will begin by recalling how to solve a linear differential equation. This can be done systematically and leads to expressing the solution as a combination of one- and two-dimensional motions, which explains why all linear systems look similar. We will then generalize the results obtained in the 2D case and discuss the qualitative behavior of the trajectories around fixed points depending on the structure of the linearized system.

Finally, we will generalize the linear stability problem to the case of a time-varying trajectory, as a prerequisite to periodic orbit stability analysis and to Lyapunov exponent characterization of chaotic dynamics.

2.1 Bidimensional phase space

To introduce the principle of a linear stability analysis in the vicinity of a fixed point we first detail all the possible cases that can be encountered in the 2D phase space.

2.1.1 Jacobian matrix

The Jacobian matrix has been defined in Section 1.3.2 when we wrote the first-order expansion of a 2D dynamical system around a fixed point (Eq. 1.20). Let us write this expansion again.

In the bidimensional case, an ODE dynamical system can be written as

$$\begin{cases} \dot{x} = f(x,y), \\ \dot{y} = g(x,y). \end{cases} \tag{2.1}$$

If we consider a small perturbation in the vicinity of a fixed point $\mathbf{X}^* = \begin{pmatrix} x^* \\ y^* \end{pmatrix}$,

$$\mathbf{X} = \mathbf{X}^* + \delta\mathbf{X} = \begin{pmatrix} x^* + \delta x \\ y^* + \delta y \end{pmatrix},$$

https://doi.org/10.1515/9783110677874-002

then we can linearize system (2.1) around $\mathbf{X}^*$:

$$\begin{cases} \frac{d}{dt}\delta x = f(x^* + \delta x, y^* + \delta y) \simeq f(x^*, y^*) + \frac{\partial f}{\partial x}\big|_{\mathbf{X}^*}\delta x + \frac{\partial f}{\partial y}\big|_{\mathbf{X}^*}\delta y, \\ \frac{d}{dt}\delta y = g(x^* + \delta x, y^* + \delta y) \simeq g(x^*, y^*) + \frac{\partial g}{\partial x}\big|_{\mathbf{X}^*}\delta x + \frac{\partial g}{\partial y}\big|_{\mathbf{X}^*}\delta y. \end{cases}$$

Using the notation $L_{ij} = \frac{\partial F_i}{\partial x_j}\big|_{\mathbf{X}^*}$, we have

$$\begin{cases} \frac{d}{dt}\delta x = L_{11}\delta x + L_{12}\delta y, \\ \frac{d}{dt}\delta y = L_{21}\delta x + L_{22}\delta y. \end{cases}$$

The general solution of this linear system depends on the nature of the eigenvalues of the Jacobian matrix

$$\mathcal{L}|_{\mathbf{X}^*} = \begin{pmatrix} L_{11} & L_{12} \\ L_{21} & L_{22} \end{pmatrix}.$$

To determine the eigenvalues of $\mathcal{L}|_{\mathbf{X}^*}$, we recall that they are the roots of the characteristic polynomial

$$p(\lambda) = \mathrm{Det}(\mathcal{L}|_{\mathbf{X}^*} - \lambda\mathbb{1}) = \begin{vmatrix} L_{11} - \lambda & L_{12} \\ L_{21} & L_{22} - \lambda \end{vmatrix}$$

$$= \lambda^2 - \mathrm{Tr}(\mathcal{L}|_{\mathbf{X}^*})\lambda + \mathrm{Det}(\mathcal{L}|_{\mathbf{X}^*}) \tag{2.2}$$

where $\mathbb{1}$ is the identity matrix. We can see that the characteristic polynomial of $\mathcal{L}|_{\mathbf{X}^*}$ and thus the eigenvalues only depend on the determinant and trace of $\mathcal{L}|_{\mathbf{X}^*}$. These are invariant properties of the underlying linear operator which are independent of the basis chosen. In particular, if $\mathcal{L}|_{\mathbf{X}^*}$ can be diagonalized, then we have $\mathrm{Det}(\mathcal{L}|_{\mathbf{X}^*}) = \lambda_1\lambda_2$ and $\mathrm{Tr}(\mathcal{L}|_{\mathbf{X}^*}) = \lambda_1 + \lambda_2$, where $\lambda_{1,2}$ are the two eigenvalues of $\mathcal{L}|_{\mathbf{X}^*}$.

We will now review all possible configurations for the eigenvalues $\lambda_{1,2}$.

2.1.2 Two distinct and nonzero real roots

The characteristic polynomial can then be factorized as $p(\lambda) = (\lambda - \lambda_1)(\lambda - \lambda_2)$, and the matrix $\mathcal{L}|_{\mathbf{X}^*}$ is diagonalizable. Any initial condition $\delta\mathbf{X}(0)$ can be decomposed in a basis formed by the eigenvectors $\mathbf{V}_1$ and $\mathbf{V}_2$ associated respectively with eigenvalues λ_1 and λ_2: $\delta\mathbf{X}(0) = a_1\mathbf{V}_1 + a_2\mathbf{V}_2$, where a_1 and a_2 are two real constants. The linear behavior with time of $\delta\mathbf{X}(t)$ is then given by $\delta\mathbf{X}(t) = a_1 e^{\lambda_1 t}\mathbf{V}_1 + a_2 e^{\lambda_2 t}\mathbf{V}_2$.

Depending on the signs of λ_1 and λ_2, we can have the following behaviors:

2.1.2.a The two roots λ_1 and λ_2 have the same sign

The fixed point is then called a *node*. The node is *stable* when the two roots are negative (see Fig. 2.1a) and is *unstable* when they are positive (see Fig. 2.1b). If the eigenvalues are significantly different, the dynamics along the most stable (or least unstable) direction quickly dies out, leaving alone the dynamics along the other eigendirection.

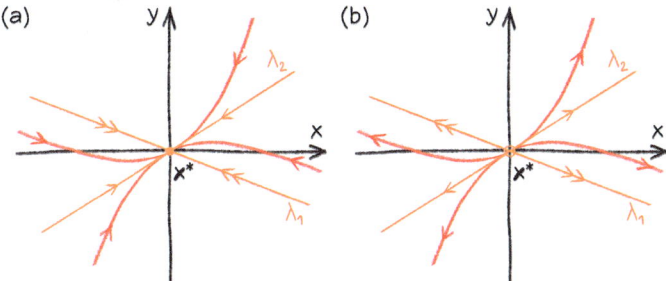

Figure 2.1: Example of a (a) *stable node* and (b) *unstable node*. The orange lines are the directions given by the eigenvectors with double arrows indicating the direction along which the dynamics is faster (i. e., $|\lambda_1| \gg |\lambda_2|$). The red curves are examples of trajectories in the phase space.

2.1.2.b The two roots λ_1 and λ_2 have opposite signs

The fixed point is called a *saddle point*. It is necessarily *unstable* since one direction is unstable. Sooner or later, the dynamics along the stable direction becomes negligible compared to that along the unstable direction (Fig. 2.2).

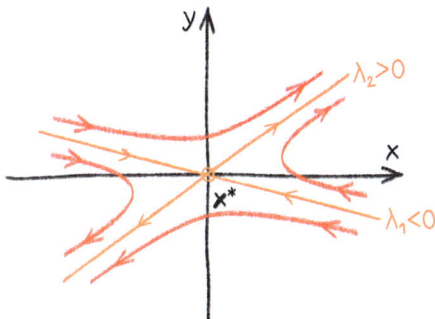

Figure 2.2: Example of a *saddle point*. The orange lines are the directions given by the eigenvectors. The red curves are examples of trajectories in the phase space.

2.1.3 Two complex conjugate roots

In this case the characteristic polynomial cannot be factorized in $\mathbb{R}$. The complex conjugate eigenvalues can be written as $\lambda_1 = \mu + i\omega$ and $\lambda_2 = \lambda_1^* = \mu - i\omega$ with $\mu, \omega \in \mathbb{R}$.

The change of coordinates $\left(\begin{smallmatrix}x\\y\end{smallmatrix}\right) \rightarrow \left(\begin{smallmatrix}x'\\y'\end{smallmatrix}\right)$ allows us to rewrite the system as (see Section 2.2.1.b):

$$\begin{cases} \dot{x}' = \mu x' + \omega y', \\ \dot{y}' = -\omega x' + \mu y'. \end{cases}$$

Then the solution of the system for an initial condition $\left(\begin{smallmatrix}x'_0\\y'_0\end{smallmatrix}\right)$ is

$$\begin{cases} x'(t) = e^{\mu t}(x'_0 \cos \omega t + y'_0 \sin \omega t), \\ y'(t) = e^{\mu t}(-x'_0 \sin \omega t + y'_0 \cos \omega t). \end{cases} \tag{2.3}$$

The $\cos \omega t$ and $\sin \omega t$ terms in (2.3) indicate the oscillatory nature of the dynamics when the imaginary part ω of an eigenvalue is nonzero. The distance to the origin is controlled by the factor $e^{\mu t}$. This indicates that it is the real part μ of the complex conjugate eigenvalues that determines the stability of the fixed point.

When $\mu \neq 0$, the fixed point at the origin is called a *spiral point* (or a *focus*) as trajectories spiral toward it or out of it (Fig. 2.3). It is stable for $\mu < 0$ and unstable for $\mu > 0$.

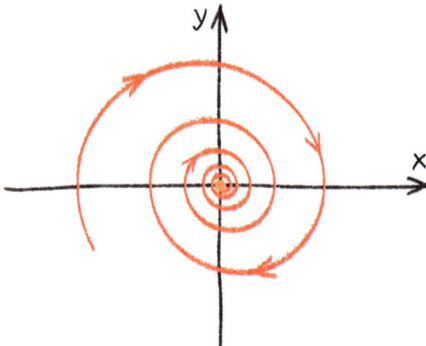

Figure 2.3: Example of a *stable spiral*. The red curve is an example of trajectory in the phase space. The unstable spiral is a similar figure with reverse arrows.

When $\mu = 0$, the stability of the fixed point cannot be determined by a linear analysis: the nonlinearities will determine if the point is actually stable or unstable. Such points are called *neutrally stable*. In this case a slight variation of a parameter value will lead to a change of stability of the fixed point, typically from a stable spiral to an unstable one. Such a change of behavior is called a *bifurcation* and will be discussed in Chapter 3.

2.1.4 Double root different from zero

This situation occurs when two roots are degenerate: $\lambda_1 = \lambda_2$. The characteristic polynomial then reads $p(\lambda) = (\lambda - \lambda_0)^2$, where λ_0 is the double root.

Care must be taken, because the eigenspace associated with λ_0 may have a dimension of 1 or 2, depending on the actual number of independent eigenvectors associated with the eigenvalue (which has an algebraic multiplicity of 2). When we solve the eigenproblem and find that there is a single eigenvector associated with a double root, the matrix $\mathcal{L}|_{\mathbf{x}^*}$ cannot be transformed into a diagonal matrix through a change of basis.

2.1.4.a $\mathcal{L}|_{\mathbf{x}^*}$ is diagonalizable

In this case, there are two eigenvectors associated with λ_0, and $\mathcal{L}|_{\mathbf{x}^*}$ can be diagonalized and is equal to $\lambda_0 \mathbb{1}$. The dynamics is isotropic. The fixed point is called a *star node*, as any line going through the origin is a trajectory (Fig. 2.4). A star node is stable (resp., unstable) if $\lambda_0 < 0$ (resp., $\lambda_0 > 0$). If $\lambda_0 = 0$, then the behavior around the fixed point is governed by nonlinear terms of higher order.

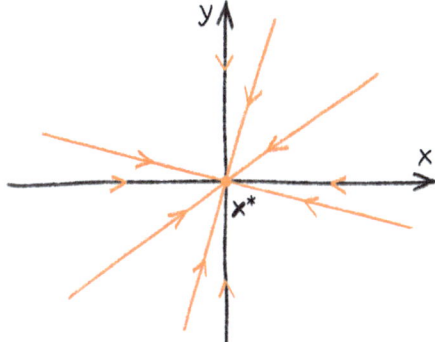

Figure 2.4: Example of a *stable star node*. All the lines going to the origin are trajectories. The figure of the unstable star node is obtained by reversing the arrows.

2.1.4.b $\mathcal{L}|_{\mathbf{x}^*}$ is not diagonalizable

There is only one eigenvector associated with λ_0, and thus a basis of eigenvectors is not available. The matrix $\mathcal{L}|_{\mathbf{x}^*}$ can nevertheless be transformed to its *Jordan form*:

$$J_2(\lambda_0) = \begin{pmatrix} \lambda_0 & 1 \\ 0 & \lambda_0 \end{pmatrix}.$$

In the new basis the equation $\dot{\mathbf{X}}' = J_2(\lambda_0)\mathbf{X}'$ can be integrated to obtain the time evolution of the coordinates (x', y') given initial condition (x'_0, y'_0):

$$\begin{cases} x'(t) = (x_0' + y_0't)e^{\lambda_0 t}, \\ y'(t) = y_0'e^{\lambda_0 t}. \end{cases}$$

Due to the degeneracy of the system, we note the appearance of a linearly growing term, called a *secular term*, which causes the trajectory to align along the unique eigenvector in the direction of x' (Fig. 2.5). Such a fixed point is called a *degenerate node*. It is stable (resp., unstable) if $\lambda_0 < 0$ (resp., $\lambda_0 > 0$). Again, if $\lambda_0 = 0$, then the behavior around the fixed point is governed by nonlinear terms.

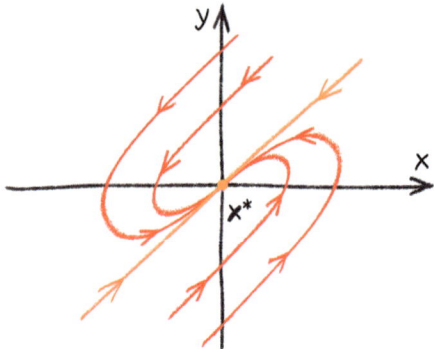

Figure 2.5: Example of a *stable degenerate node*. The orange line corresponds to the unique eigendirection. The red curves are examples of trajectories in the phase space. A portrait of an unstable degenerate node is obtained by reversing the arrows.

2.1.5 Phase diagram in terms of the trace and determinant of the Jacobian matrix

In the introduction of Section 2.1.1, we noted that the eigenvalues of the Jacobian matrix can be computed from its trace and determinant, denoted T and Δ, respectively. Recalling that the characteristic polynomial of the Jacobian matrix is (see Eq. (2.2))

$$p(\lambda) = \lambda^2 - T\lambda + \Delta,$$

the eigenvalues are given by

$$\lambda_\pm = \frac{T}{2} \pm \sqrt{\frac{T^2}{4} - \Delta} \quad \text{when} \quad \frac{T^2}{4} > \Delta,$$

$$\lambda_\pm = \frac{T}{2} \pm i\sqrt{\Delta - \frac{T^2}{4}} \quad \text{when} \quad \frac{T^2}{4} < \Delta.$$

We can thus deduce the following general properties:
- When $\Delta = \lambda_+\lambda_- < 0$, the two eigenvalues are necessarily real (since $\frac{T^2}{4} > \Delta$) and have opposite signs, so that we have a saddle point.

- When $\Delta > 0$ and $\frac{T^2}{4} > \Delta$, the two eigenvalues are real and have the same sign. The fixed point is a node. Since $|T|/2 > \sqrt{T^2/4 - \Delta}$, it is stable (resp., unstable) when $T < 0$ (resp., $T > 0$), the two eigenvalues being negative (resp., positive).
- When $\Delta > 0$ and $\frac{T^2}{4} < \Delta$, the two eigenvalues are complex conjugates, and we have a spiral point. The real part of the eigenvalues is $\frac{T}{2}$, and thus the spiral point is stable (resp., unstable) when $T < 0$ (resp., $T > 0$).

Interestingly, the stability condition can be globally written as $\Delta > 0$, $T < 0$. The curve $\Delta = \frac{T^2}{4}$ separates spiral points from nodes. This analysis is summarized in Fig. 2.6 (Strogatz, 2018).

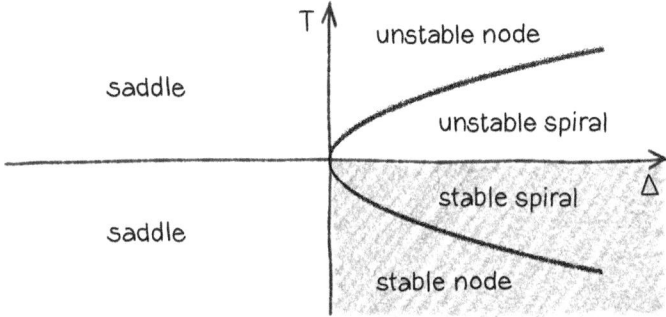

Figure 2.6: Summary of all possible cases in 2 dimensions in terms of the values of the determinant and trace of the matrix. The parabola separating spiral from node points has equation $\Delta = T^2/4$. The hatched part of the plane corresponds to the stable fixed points.

2.1.6 Applications

2.1.6.a Application to the damped pendulum
The dynamical system describing the damped pendulum is (see Section 1.1.3):

$$\begin{cases} \dot{X}_1 = X_2, \\ \dot{X}_2 = -\frac{g}{l} \sin X_1 - \gamma X_2, \end{cases}$$

where $\gamma > 0$ and $g/l > 0$. To find the fixed points, we search the points $\mathbf{X}^*$ that are solutions of

$$\begin{cases} 0 = X_2^*, \\ 0 = -\frac{g}{l} \sin X_1^* - \gamma X_2^*. \end{cases}$$

We deduce that the points $(0,0)$ and $(\pi, 0)$ are fixed points as well as all points $(2n\pi, 0)$ and $(\pi + 2n\pi, 0)$, $n \in \mathbb{Z}$, due to the 2π-periodicity.

To study the stability of those two points, we first compute the Jacobian matrix of the system:

$$\mathcal{L} = \begin{pmatrix} 0 & 1 \\ -\frac{g}{l}\cos X_1 & -\gamma \end{pmatrix}.$$

Then we compute the Jacobian matrix at the fixed points.

2.1.6.a-i Stability of (0, 0)

Since $X_1 = 0$, we have

$$\mathcal{L}|_{(0,0)} = \begin{pmatrix} 0 & 1 \\ -\frac{g}{l} & -\gamma \end{pmatrix},$$

$T = -\gamma < 0$, and $\Delta = \frac{g}{l}$. Consequently, the fixed point $(0,0)$ is stable. To determine if it is a node or a spiral, we compare Δ to $T^2/4$:

- If $\frac{\gamma^2}{4} > \frac{g}{l}$, then the fixed point is a node (overdamped case).
- If $\frac{\gamma^2}{4} < \frac{g}{l}$, then the fixed point is a spiral (underdamped case).

In the particular case $\gamma = 0$ the trajectories do not spiral toward or away from the fixed point. In this example, the linear part does not determine the stability of the fixed point. This is because without damping the pendulum is a conservative system. In this case, and for conservative systems, the fixed point is called a *center*.

2.1.6.a-ii Stability of $(\pi, 0)$

Here, $X1 = \pi$ so that

$$\mathcal{L}|_{(\pi,0)} = \begin{pmatrix} 0 & 1 \\ \frac{g}{l} & -\gamma \end{pmatrix},$$

with $T = -\gamma < 0$, and $\Delta = -\frac{g}{l}$. Consequently, the fixed point $(\pi, 0)$ is a saddle point, which is always unstable.

2.1.6.b Model of the chlorine dioxide-iodine-malonic acid chemical reaction

This second exemple is analyzed in (Strogatz, 2018). It is a very simplified model of the ClO_2-I_2-$CH_2(COOH)_2$ chemical reaction:

$$\frac{dx}{dt} = a - x - \frac{4xy}{1 + x^2},$$
$$\frac{dy}{dt} = bx\left(1 - \frac{y}{1 + x^2}\right),$$

where a and b are positive parameters, and x and y are positive variables proportional to the concentrations of the reactants.

This system has a unique fixed point $(x^*, y^*) = (\frac{a}{5}, 1 + \frac{a^2}{25})$. The Jacobian of the system is

$$\mathcal{L} = \begin{pmatrix} -1 + 4y\frac{x^2-1}{(1+x^2)^2} & -\frac{4x}{1+x^2} \\ b(1 + y\frac{x^2-1}{(1+x^2)^2}) & -\frac{bx}{1+x^2} \end{pmatrix},$$

and its expression at the fixed point (x^*, y^*) is

$$\mathcal{L}|_{(x^*, y^*)} = \frac{1}{1+x^{*2}} \begin{pmatrix} 3x^{*2} - 5 & -4x^* \\ 2bx^{*2} & -bx^* \end{pmatrix}.$$

The determinant of this matrix is $\Delta = 5bx^*/(1 + x^{*2})$, which is always positive as x^* is a concentration and is thus positive. The trace is

$$T = \frac{3x^{*2} - bx^* - 5}{1 + x^{*2}},$$

and its sign depends on the values of a and b. Replacing x^* by its expression, we find that $3x^{*2} - bx^* - 5 = \frac{3a^2}{25} - \frac{ab}{5} - 5 < 0$ if $b > \frac{3a}{5} - \frac{25}{a}$.

Consequently, the fixed point is a stable spiral for $b > \frac{3a}{5} - \frac{25}{a}$ and an unstable one for $b < \frac{3a}{5} - \frac{25}{a}$. In the second case the asymptotic behavior is in fact oscillatory, as we will see in Chapter 4.

2.2 Linear stability analysis of a fixed point in arbitrary dimension

In this section, we consider the general problem of the stability of a fixed point in any dimension. Since a stability analysis involves solving linear differential equations, we begin by reviewing how to solve them generally in Section 2.2.1. Then we formulate the stability analysis problem in Section 2.2.2. Finally, we summarize our results in Section 2.2.3, where we discuss the relation between the spectrum of eigenvalues of the Jacobian matrix and the stability of the fixed point. This allows us to anticipate situations where a change of stability induces qualitative changes in the dynamics, situations called bifurcations, which we will study in Chapters 3 and 4.

2.2.1 Linear differential equations in arbitrary dimension

Consider a system of first-order linear differential equations with constant coefficients, written as

$$\frac{dX}{dt} = M \cdot X \qquad (2.4)$$

with an $n \times n$ matrix M with constant coefficients in $\mathbb{R}$ and $X \in \mathbb{R}^n$. The solutions of this equation are given by

$$X(t) = e^{tM} \cdot X(0), \qquad (2.5)$$

where the matrix exponential e^A is defined for any square matrix A by the convergent series

$$e^A = \mathbb{1} + A + \frac{A^2}{2!} + \frac{A^3}{3!} + \cdots + \frac{A^p}{p!} + \cdots. \qquad (2.6)$$

The matrix exponentials e^{tM} have the following properties:

$$e^{t_1 M} e^{t_2 M} = e^{(t_1 + t_2)M}, \qquad (2.7a)$$

$$\frac{d}{dt} e^{tM} = M e^{tM}, \qquad (2.7b)$$

$$\mathrm{Det}(e^{tM}) = e^{t\,\mathrm{Tr}(M)}. \qquad (2.7c)$$

as can be verified using definition (2.6). Relation (2.7b) ensures that (2.5) satisfies Eq. (2.4).

2.2.1.a Case of a diagonal matrix

Even if it is a particular case, we get a sense of how the matrix exponential behaves by assuming that M is a diagonal matrix

$$M = \begin{pmatrix} \lambda_1 & & 0 \\ & \ddots & \\ 0 & & \lambda_n \end{pmatrix}$$

with eigenvalues $\lambda_i \in \mathbb{R}$.

The exponential matrix e^{tM} is then given by

$$e^{tM} = \begin{pmatrix} e^{\lambda_1 t} & & 0 \\ & \ddots & \\ 0 & & e^{\lambda_n t} \end{pmatrix}.$$

Choosing as an initial condition the normalized eigenvector of M associated with the eigenvalue λ_i,

$$\mathbf{X_0} = \mathbf{V}_i = \begin{pmatrix} 0 \\ \vdots \\ 0 \\ 1 \\ 0 \\ \vdots \\ 0 \end{pmatrix} \cdots ith,$$

the trajectory is given by

$$\mathbf{X}(t) = e^{t\mathbf{M}} \cdot \mathbf{X_0} = e^{\lambda_i t}\mathbf{X_0} = \begin{pmatrix} 0 \\ \vdots \\ e^{\lambda_i t} \\ \vdots \\ 0 \end{pmatrix}.$$

Depending on the value of λ_i, three different asymptotic behaviors are possible:

- if $\lambda_i < 0$, then $\lim_{t \to +\infty} e^{\lambda_i t} = 0$, and $\mathbf{X}(t)$ converges to the origin;
- if $\lambda_i = 0$, then $\mathbf{X}(t)$ remains equal to $\mathbf{X_0}$;
- if $\lambda_i > 0$, then $\lim_{t \to +\infty} e^{\lambda_i t} = \infty$, and $\mathbf{X}(t)$ explodes to infinity.

Considering now an arbitrary initial condition $\mathbf{X_0}$, we will decompose the initial condition $\mathbf{X_0}$ on the basis of eigenvectors $\mathbf{V_i}$ associated with the eigenvalues λ_i:

$$\mathbf{X_0} = \sum_i a_i \mathbf{V_i}.$$

The time evolution of $\mathbf{X}(t)$ is then given by

$$\mathbf{X}(t) = \sum_i a_i e^{\lambda_i t} \mathbf{V_i}. \tag{2.8}$$

We see that the only case where $\mathbf{X_0}(t) \to 0$ is when all eigenvalues satisfy $\lambda_i < 0$. It suffices for a single value to be strictly positive for the corresponding term to increase exponentially with time, driving the state vector to infinity.

2.2.1.b General case

Let us now assume that $\mathbf{M}$ is not diagonal. In this case, the solution is to find a change of basis that makes the matrix as diagonal as possible.

If $\mathbf{M}$ is diagonalizable, then we can choose a new basis consisting of eigenvectors. It is often the case but not always. Let us first assume that this is the case. Assuming that the new vector coordinates are given by $\mathbf{Y} = \mathbf{PX}$, the matrix $\mathbf{D}$ representing in the new basis the linear operator associated with $\mathbf{M}$ in the old basis is given by

$$\mathbf{D} = \mathbf{PMP}^{-1},$$

so that $\mathbf{DPX}$ (change the basis, then apply the operator) and $\mathbf{PMX}$ (apply the operator, then change the basis) are identical. The matrices $\mathbf{M}$ and $\mathbf{D}$ representing the same linear operator in different bases have the same set of eigenvalues. Then we can show using (2.6) that the matrix exponentials follow the same transformation:

$$e^{t\mathbf{D}} = \mathbf{P}e^{t\mathbf{M}}\mathbf{P}^{-1},$$

so that the time evolution of a state vector in the new basis is given by

$$\mathbf{Y}(t) = e^{t\mathbf{D}} \cdot \mathbf{Y}(0),$$

and so we are back to the case of a diagonal matrix.

What happens when the matrix is not diagonalizable and what does it mean for a matrix to be as diagonal as possible? To answer this question, we have to consider the eigenvalues of $\mathbf{M}$, as well as their algebraic and geometric multiplicities.

Recall that an eigenvalue of $\mathbf{M}$ is a number λ such that $p_{\mathbf{M}}(\lambda) = \mathrm{Det}(\mathbf{M} - \lambda\mathbb{1}) = 0$. If $\mathbf{M}$ is a real matrix, then its eigenvalues can be real numbers or pairs of complex conjugate numbers. The algebraic multiplicity of λ is its multiplicity as a root of the characteristic polynomial $p_{\mathbf{M}}(\lambda)$. Its geometric multiplicity is the dimension of the vector subspace associated with the eigenvalue, called an eigenspace, and is given by $n - r$ where n is the dimension of the matrix and r is the rank of $(\mathbf{M} - \lambda\mathbb{1})$. It may be that the geometric multiplicity of an eigenvalue is strictly smaller than its algebraic multiplicity, and then the matrix is not diagonalizable.

In general, $\mathbf{M}$ can be brought to a block-diagonal form

$$\mathbf{M} = \mathbf{P}^{-1} \begin{pmatrix} (\mathbf{B}_1) & 0 & 0 & 0 \\ 0 & (\mathbf{B}_2) & 0 & 0 \\ 0 & 0 & (\mathbf{B}_3) & 0 \\ 0 & 0 & 0 & \ddots \end{pmatrix} \mathbf{P},$$

where $(\mathbf{B}_i)$ are real numbers or square matrices depending on the nature of the associated eigenvalue. In particular,

- Any real eigenvalue λ_i whose algebraic multiplicity r (typically 1) equals its geometric multiplicity is associated with r diagonal terms λ_i, yielding r diagonal terms $e^{\lambda_i t}$ in the matrix exponential.
- Any pair of complex conjugate eigenvalues $\mu_i \pm i\omega_i$ is associated with a 2×2 block

$$C(\mu_i, \omega_i) = \begin{pmatrix} \mu_i & \omega_i \\ -\omega_i & \mu_i \end{pmatrix} \quad \text{with } e^{C(\mu_i, \omega_i)t} = \begin{pmatrix} e^{\mu_i t} \cos \omega_i t & e^{\mu_i t} \sin \omega_i t \\ -e^{\mu_i t} \sin \omega_i t & e^{\mu_i t} \cos \omega_i t \end{pmatrix}.$$

– Any real eigenvalue λ_i with algebraic multiplicity r and geometric multiplicity 1 is associated with a so-called $r \times r$ Jordan block

$$J_r(\lambda_i) = \begin{pmatrix} \lambda_i & 1 & 0 & \cdots \\ 0 & \lambda_i & 1 & \\ 0 & 0 & \lambda_i & \ddots \\ 0 & 0 & 0 & \ddots \end{pmatrix} \quad \text{with } e^{J_r(\lambda_i)t} = \begin{pmatrix} e^{\lambda_i t} & te^{\lambda_i t} & \frac{t^2}{2}e^{\lambda_i t} & \cdots \\ 0 & e^{\lambda_i t} & te^{\lambda_i t} & \ddots \\ 0 & 0 & e^{\lambda_i t} & \ddots \\ 0 & 0 & 0 & \ddots \end{pmatrix}$$

with λ_i on the diagonal and 1s above the diagonal.

Now

$$e^{\mathbf{M}t} = \mathbf{P}^{-1} \begin{pmatrix} e^{\mathbf{B}_1 t} & 0 & 0 & 0 \\ 0 & e^{\mathbf{B}_2 t} & 0 & 0 \\ 0 & 0 & e^{\mathbf{B}_3 t} & 0 \\ 0 & 0 & 0 & \ddots \end{pmatrix} \mathbf{P}.$$

We have omitted the rather exceptional and unusual case where the algebraic and geometric multiplicities of a *complex* eigenvalue do not agree (in which case we can always pick a complex change of basis to obtain a complex block of the form $J_r(\lambda)$).

When the matrix $\mathbf{M}$ features Jordan blocks or complex eigenvalue blocks, the eigenvectors of $\mathbf{M}$ no longer form a basis because (1) the eigenvectors of the complex eigenvalue blocks are complex and only some of their linear combinations are real and (2) the eigenvector associated with a Jordan block spans a vector space of dimension lower than the algebraic multiplicity of the eigenvalue.

However, we can see that each eigenspace associated with a given block is invariant: if the initial condition is contained within this subspace, then the dynamics will remain confined to it. Moreover, the matrix elements of each type of diagonal block converge to zero whenever the associated eigenvalue (when it is real) or its real part (when it is complex) is negative, and thus the dynamics in that vector subspace vanishes exponentially fast.

This confirms the conclusion of Section 2.2.1.a in the general case of a nondiagonal matrix $\mathbf{M}$: (1) if all eigenvalues have a negative real part, then any state vector will converge to zero; (2) if some eigenvalue has a positive real part, then the dynamics in the corresponding eigenspace explodes to infinity.

2.2.2 Linear stability analysis

Recall that the *fixed points* of a system are the particular points $\mathbf{X}^*$ of the phase space that satisfy

$$\frac{d\mathbf{X}}{dt}\bigg|_{\mathbf{X}^*} = 0 \quad \text{or} \quad \mathbf{F}(\mathbf{X}^*) = 0.$$

If the system is located at a fixed point, then it will remain there forever. Fixed points are thus the simplest invariant sets. Fixed points can be stable or unstable. Only fully stable fixed points are attractors.

To study how the trajectories behave in a neighborhood of a fixed point $\mathbf{X}^*$, we linearize the equations around this point. Writing $\mathbf{X}$ as perturbation of $\mathbf{X}^*$,

$$\mathbf{X} = \mathbf{X}^* + \delta\mathbf{X} \quad \text{with } \delta\mathbf{X} \text{ small,}$$

we have:

$$\frac{d\mathbf{X}}{dt} = \mathbf{F}(\mathbf{X}),$$

$$\frac{d\mathbf{X}^*}{dt} + \frac{d(\delta\mathbf{X})}{dt} = \mathbf{F}(\mathbf{X}^* + \delta\mathbf{X}).$$

Let us decompose the last equation into component equations (see Eq. 1.4):

$$\begin{cases} \frac{d(\delta X_1)}{dt} = F_1(X_1^* + \delta X_1, X_2^* + \delta X_2, \ldots, X_n^* + \delta X_n), \\ \frac{d(\delta X_2)}{dt} = F_2(X_1^* + \delta X_1, X_2^* + \delta X_2, \ldots, X_n^* + \delta X_n), \\ \quad \vdots \\ \quad \vdots \\ \frac{d(\delta X_n)}{dt} = F_n(X_1^* + \delta X_1, X_2^* + \delta X_2, \ldots, X_n^* + \delta X_n). \end{cases} \tag{2.9}$$

Computing the Taylor expansion of each function, we obtain

$$\begin{cases} \frac{d(\delta X_1)}{dt} = F_1(\mathbf{X}^*) + \frac{\partial F_1}{\partial X_1}\big|_{\mathbf{X}^*} \delta X_1 + \frac{\partial F_1}{\partial X_2}\big|_{\mathbf{X}^*} \delta X_2 + \cdots + \frac{\partial F_1}{\partial X_n}\big|_{\mathbf{X}^*} \delta X_n, \\ \quad \vdots \\ \frac{d(\delta X_n)}{dt} = F_n(\mathbf{X}^*) + \frac{\partial F_n}{\partial X_1}\big|_{\mathbf{X}^*} \delta X_1 + \frac{\partial F_n}{\partial X_2}\big|_{\mathbf{X}^*} \delta X_2 + \cdots + \frac{\partial F_n}{\partial X_n}\big|_{\mathbf{X}^*} \delta X_n. \end{cases} \tag{2.10}$$

A condensed way to write this system is to use the Jacobian matrix of $\mathbf{F}$ with elements

$$\mathcal{L}_{ij} = \frac{\partial F_i}{\partial X_j}.$$

Then Eq. (2.10) can be written as

$$\frac{d(\delta\mathbf{X})}{dt} \simeq \mathbf{F}(\mathbf{X}^*) + \mathcal{L}|_{\mathbf{X}^*} \cdot \delta\mathbf{X},$$

and since $\mathbf{F}(\mathbf{X}^*) = 0$, we obtain

$$\frac{d(\delta\mathbf{X})}{dt} = \mathcal{L}|_{\mathbf{X}^*} \cdot \delta\mathbf{X}. \tag{2.11}$$

As $\mathcal{L}|_{\mathbf{X}^*}$ is a matrix with constant coefficients, Eq. (2.11) is a linear ordinary differential equation, which can be explicitly integrated as discussed in Section 2.2.1, its solutions being of the form $\delta\mathbf{X}(t) = e^{(t\mathcal{L}|_{\mathbf{X}^*})}\delta\mathbf{X}(0)$.

We thus have the following properties:

- If an initial condition $\delta\mathbf{X}(0)$ is aligned along an eigenvector of $\mathcal{L}|_{\mathbf{X}^*}$ or is contained in a vector subspace associated with an eigenvalue λ_+ of *strictly positive real part*, then $\delta\mathbf{X}(t)$ remains in that subspace, and $\|\delta\mathbf{X}(t)\| \to \infty$, that is, the system moves away from $\mathbf{X}_0$. These eigenvectors or vector subspaces correspond to *unstable or dilatant directions*.

- If an initial condition $\delta\mathbf{X}(0)$ is aligned along an eigenvector of $\mathcal{L}|_{\mathbf{X}^*}$ or is contained in a vector subspace associated with an eigenvalue λ_- of *strictly negative real part*, then $\delta\mathbf{X}(t)$ remains in that subspace, and $\|\delta\mathbf{X}(t)\| \to 0$, that is, the system converges to $\mathbf{X}_0$. These eigenvectors or vector subspaces correspond to *stable or contracting directions*.

- In the general case, if there exists an unstable eigenspace, and if the initial condition $\delta\mathbf{X}_0$ has a non-zero component in it, then the system will diverge away from $\mathbf{X}_0$, otherwise it will converge to it.

- The eigenvalues with zero real part are problematic because the linear approximation is no longer valid due to the vanishing of the corresponding terms. We then have to continue the Taylor expansion to include the first nonvanishing terms.

Consequently, the fixed point $\mathbf{X}_0$ attracts all neighboring trajectories if all eigenvalues of the Jacobian matrix have negative real parts.

2.2.3 The spectrum of the Jacobian matrix determines the stability of a fixed point

In conclusion, the stability of a fixed point depends on the set of eigenvalues of the Jacobian matrix at this point (i. e., its spectrum). If all these eigenvalues have negative real parts (Fig. 2.7a), then the fixed point is stable.

We will see in the next chapter that the eigenvalues typically change when a parameter varies. An important question then is whether one or several eigenvalues can cross the imaginary axis in Fig. 2.7 because their real parts would then change sign, modifying the stability. Typically, such a crossing can occur in two different ways:

- a single real eigenvalue crosses the origin (Fig. 2.7b). Chapter 3 will deal with this case;

- a pair of complex conjugate eigenvalues simultaneously crosses the imaginary axis (Fig. 2.7c). This case will be discussed in Chapter 4.

Note that some fixed points may display several eigenvalues with positive real parts, having experienced multiple bifurcations. These fixed points have then several unstable directions or planes.

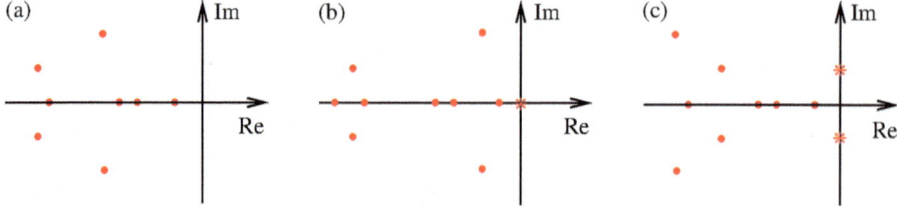

Figure 2.7: Representation of the set of the eigenvalues of a Jacobian matrix at a fixed point. (a) Stable case. (b) A single eigenvalue has a zero real part. (c) A pair of complex conjugate eigenvalues has a zero real part.

2.3 From linear eigenspaces to invariant manifolds

The linear stability analysis of a dynamical system around a fixed point provides us with eigenvalues and their associated eigenspaces. We have seen that eigenspaces govern trajectories around the fixed point; however, they are only valid in a neighborhood of it. Here we discuss how eigenspaces extend into invariant manifolds in the entire phase space, elaborating on Section 1.3.2.b.

Without loss of generality, we can consider a fixed point located at the origin. In the case of a linear flow specified by $\dot{\mathbf{X}} = \mathbf{L}\mathbf{X}$, the eigenvectors of the Jacobian are relevant to the dynamics in the entire space:

- For a real eigenvalue, the straight line passing through the origin in the direction of the eigenvector is an invariant set since the velocity vector is aligned along the line (Fig. 2.8a).
- For a pair of complex conjugate eigenvalues, the plane defined by the two directions associated with this pair also forms an invariant set, where asymptotic behavior is governed by the common real part of these eigenvalues.

In both cases, if the real part of the eigenvalue is negative (resp., positive), then the orbit of a point located along the line or in the plane will converge to the fixed point as $t \to \infty$ (resp., $t \to -\infty$) (Fig. 2.8a).

If the system is nonlinear and is defined by $\dot{\mathbf{X}} = \mathbf{L}\mathbf{X} + \mathbf{G}(\mathbf{X})$ where $\mathbf{G}(\mathbf{X})$ is the sufficiently smooth nonlinear part (satisfying $\mathbf{G}(0) = 0$, $(\partial\mathbf{G}/\partial\mathbf{X})(0) = 0$), consider the continuous deformation $\dot{\mathbf{X}} = \mathbf{L}\mathbf{X} + \xi\mathbf{G}(\mathbf{X})$ taking the linearized system around the origin to the full nonlinear system when ξ is increased from 0 to 1. Trajectories in the plane are themselves continuously deformed, in particular, those that are contained in an invariant set. Therefore the straight lines that are invariant when $\xi = 0$ deform into invariant smooth curves as ξ increases to 1 (Fig. 2.8b). Similarly, the orbits living inside an invariant plane when $\xi = 0$ will deform to trajectories contained in an invariant smooth surface.

At the origin, where the nonlinear part is negligible, these invariant curves or surfaces will be tangent to the corresponding eigenspaces. Therefore we conclude that any eigendirection (real eigenvalue) or eigenplane (a pair of complex conjugate eigenvalues) of the Jacobian matrix extends into an invariant curve or surface that is tangent

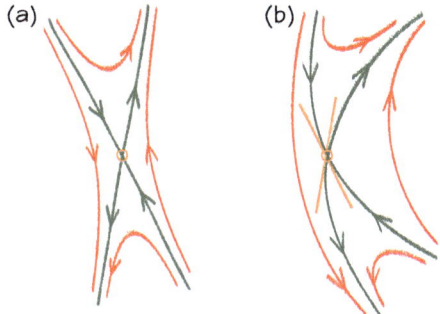

Figure 2.8: (a) In a linear system the eigendirections of the Jacobian matrix define invariant sets; (b) in a nonlinear system the eigendirections at the origin define the local direction of invariant sets that extend into the entire phase space.

to it at the origin. These invariant sets can also be constructed as follows: take an infinitesimal neighborhood V_i of the fixed point along an eigenvector or an eigenplane, and determine

$$W_i = \cup_{t>0} \phi_{\pm t}(V_i),$$

where the plus or minus sign is taken according to whether the real part of the associated eigenvalue is positive or negative. By construction the W_i are invariant sets (curves or surfaces).

Now assume that the eigenvalues satisfy

$$\mathfrak{R}(\lambda_1) \geq \mathfrak{R}(\lambda_2) \geq \cdots \geq \mathfrak{R}(\lambda_u) \geq 0 \geq \mathfrak{R}(\lambda_{u+1}) \geq \cdots \geq \mathfrak{R}(\lambda_{u+s}),$$

so that there are u (resp., s) eigenvalues with positive (resp., negative) real parts.

Together, u unstable curves or surfaces are embedded in the unstable manifold $W^u(\mathbf{X}^*)$, whereas s stable curves or surfaces are embedded in the stable manifold $W^s(\mathbf{X}^*)$. We recall the definition of stable and unstable manifolds given in Eq. (1.23):

$$W^s(\mathbf{X}^*) = \left\{ \mathbf{X} \in \mathcal{S} : \lim_{\tau \to \infty} \phi_\tau(\mathbf{X}) = \mathbf{X}^* \right\},$$
$$W^u(\mathbf{X}^*) = \left\{ \mathbf{X} \in \mathcal{S} : \lim_{\tau \to -\infty} \phi_\tau(\mathbf{X}) = \mathbf{X}^* \right\}.$$

2.4 Variational analysis around a trajectory

In this chapter, we studied the linear stability of fixed points taking advantage of the fact that the state variables for those invariant sets and hence the Jacobian matrix are constant. However, there are many other cases where we need to study the time evolution of perturbations around a solution that evolves in time. For example, in the case of periodic orbits, we need to determine the stability of these orbits.

The problem is as follows: we have a reference solution $\mathbf{X}_{\text{ref}}(t)$ of the standard equations $\dot{\mathbf{X}} = \mathbf{F}(\mathbf{X})$, and we want to study how a neighboring trajectory $\mathbf{X}(t) = \mathbf{X}_{\text{ref}}(t) + \delta\mathbf{X}(t)$ behaves compared to this reference solution (see Fig. 2.9) and to determine the asymptotic behavior of $\delta\mathbf{X}(t)$. Along the same lines as previously, we obtain that

$$\frac{\mathrm{d}\delta\mathbf{X}(t)}{\mathrm{d}t} = \left(\frac{\partial \mathbf{F}}{\partial \mathbf{X}}\right)\bigg|_{\mathbf{X}=\mathbf{X}_{\text{ref}}(t)} \delta\mathbf{X}(t) \tag{2.12}$$

with the added difficulty that the Jacobian matrix is now dependent on time. Equation (2.12) is known as the variational equation around the solution $\mathbf{X}_{\text{ref}}(t)$. Except in specific cases, it is not possible to give closed-form expressions of the solutions of Eq. (2.12), and we have to resort to numerical integration.

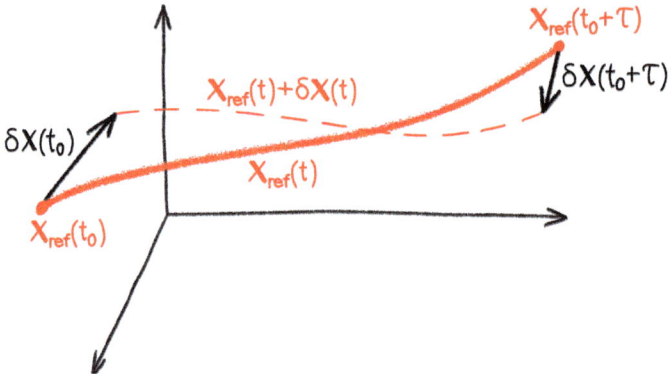

Figure 2.9: Reference trajectory $\mathbf{X}_{\text{ref}}(t)$ (solid red line) and a neighboring one $\mathbf{X}_{\text{ref}}(t) + \delta\mathbf{X}(t)$ (dashed red line).

Importantly, Eq. (2.12) takes into account a perturbation arising in one direction only, and solving it repeatedly for different initial values of the perturbation would not be very efficient. Thus we usually consider a matrix $M(t)$ whose columns are an independent set of perturbation vectors, which are initially given by the basis vectors in the tangent space. Then we numerically solve the following equation together with the original nonlinear equations $\dot{\mathbf{X}} = \mathbf{F}(\mathbf{X})$, since we need to know $(\partial \mathbf{F}/\partial \mathbf{X})$:

$$\frac{\mathrm{d}}{\mathrm{d}t}M(t) = \left(\frac{\partial \mathbf{F}}{\partial \mathbf{X}}\right)\bigg|_{\mathbf{X}=\mathbf{X}_{\text{ref}}(t)} M(t), \quad M(t_0) = \mathbb{1}, \tag{2.13}$$

where $\mathbb{1}$ is the identity matrix. The matrix $M(t)$ is known as the *fundamental matrix solution* of the linearized equation (2.12).

Following $M(t)$ along the orbit, we obtain directly the time evolution of a perturbation initially aligned along each of the directions of the tangent space. Since Eq. (2.13) is linear, the solution at time t starting from an arbitrary initial condition is a linear combination of these particular solutions, and we have that

$$\delta\mathbf{X}(t_0 + \tau) = \mathbf{M}(\tau)\delta\mathbf{X}(t_0), \tag{2.14}$$

expressing the fact that the perturbations at time $t_0 + \tau$ depend linearly on the initial perturbations. The asymptotic behavior will depend on whether there are solutions $\delta\mathbf{X}(t)$ that grow without bounds or if they all converge to zero.

Note that since the matrix $M(\tau)$ transforms the perturbation $\delta\mathbf{X}(t_0)$ around $\mathbf{X}(t_0)$ into the perturbation $\delta\mathbf{X}(t_0 + \tau)$ around $\mathbf{X}(t_0 + \tau)$, it can be considered as the Jacobian (or tangent map) of the flow ϕ_τ:

$$\mathbf{M}(\tau) = \left(\frac{\partial\phi_\tau}{\partial\mathbf{X}}\right)\Big|_{\mathbf{X}=\mathbf{X}(t_0)} = \left(\frac{\partial\mathbf{X}(t_0 + \tau)}{\partial\mathbf{X}(t_0)}\right).$$

Two types of variational analysis are of particular interest:
- the analysis of the stability of a periodic orbit of period T (Chapter 4), determined by the eigenvalues of the matrix $M(T)$, called the Floquet matrix, such that

$$\delta\mathbf{X}(t_0 + T) = \mathbf{M}(T)\delta\mathbf{X}(t_0).$$

- the characterization of chaotic dynamics, for which perturbations are growing exponentially in some directions while they shrink in others, as we will see in Sections 5.3.3 and 6.1.1.

2.5 Conclusions

In this chapter, we have studied how to perform a linear stability analysis in the neighborhood of fixed points, and we have identified general rules to assert their nature (stable or unstable). The exhaustive study of all two-dimensional cases allowed us to describe the behavior of trajectories around a fixed point in a plane. Moreover, this study helped us to understand the behavior of trajectories in arbitrary dimensions. Finally, we introduced the variational analysis of trajectories, which will be essential in the following chapters to analyze the stability of trajectories around invariant sets more complex than fixed points.

Exercises

Linear systems

System 1
Consider the system

$$\begin{cases} \dot{x} = 2x - y, \\ \dot{y} = 2y - x, \end{cases}$$

with initial conditions $x(0) = 1$ and $y(0) = 3$.

1. Show that the solution of the system is

$$x(t) = 2e^t - e^{3t},$$
$$y(t) = 2e^t + e^{3t}.$$

The system can also be written as

$$\begin{pmatrix} \dot{x} \\ \dot{y} \end{pmatrix} = \begin{pmatrix} 2 & -1 \\ -1 & 2 \end{pmatrix} \begin{pmatrix} x \\ y \end{pmatrix}.$$

2. Show that the eigenvalues of the system are 1 and 3.

System 2
Consider the system

$$\begin{cases} \dot{x} = \omega y, \\ \dot{y} = -\omega x, \end{cases}$$

with initial conditions $x(0) = 1$ and $y(0) = 7$.

1. Show that the solution is

$$x(t) = \cos \omega t + 7 \sin \omega t,$$
$$y(t) = 7 \cos \omega t - \sin \omega t.$$

The system can also be written as

$$\begin{pmatrix} \dot{x} \\ \dot{y} \end{pmatrix} = \begin{pmatrix} 0 & \omega \\ -\omega & 0 \end{pmatrix} \begin{pmatrix} x \\ y \end{pmatrix}.$$

2. Show that the eigenvalues of the system are the complex conjugates $i\omega$ and $-i\omega$.

RLC oscillator
Consider the following electric circuit:

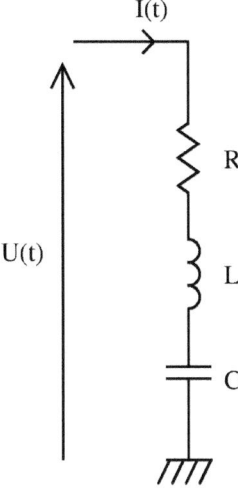

1. Show that the equation describing the dynamics of the system is

$$L\ddot{I} + R\dot{I} + \frac{I}{C} = 0.\qquad(2.15)$$

2. Rewrite Eq. (2.15) as a dynamical system of the form $\frac{dX}{dt} = F(X)$.
3. What are the characteristics of this dynamical system (dimension, conservative/dissipative, autonomous/nonautonomous)?
4. What is the Jacobian matrix of the system?
5. Draw the nullclines in the phase space and the vector field along those lines.
6. Draw the vector field in the different areas of the phase space delimited by the nullclines.
7. What are the fixed points of the system? Can you comment on their stability?
8. Study the nature of the fixed point in the two cases: $R^2C > 4L$ and $R^2C < 4L$. Draw typical trajectories in the phase space in each case.

Neutrally stable example

Consider the following system with $a \in \mathbb{R}^*$:

$$\begin{cases} \dot{x} = -y + ax(x^2 + y^2), \\ \dot{y} = x + ay(x^2 + y^2). \end{cases}$$

1. What are the fixed points of the system?

2. Compute the Jacobian of the system.
3. Perform a linear stability analysis of the fixed point. Comment.

In this example, we can determine the stability of the fixed point without linearizing the system.

Consider the following change of variable $z = x + iy$.

4. Reformulate the system in the following way:

$$\dot{z} = iz + az|z|^2.$$

5. Using the module and phase of $z = re^{i\theta}$, find the system of equations governing the dynamics of (r, θ).
6. Depending on the sign of a, determine the stability of the fixed point.

Analytical vs. graphical method

Consider the following bidimensional system:

$$\begin{cases} \dot{x} = y - x, \\ \dot{y} = x^2 - y - \mu, \end{cases}$$

with $\mu > 0$.
1. Determine the fixed points of the system.
2. Compute the Jacobian of the system.
3. Study the stability of the fixed points and deduce their nature.
4. Plot the nullclines of the system in the phase space.
5. Draw the vector field in the different areas of the phase plane delimited by the null-clines.
6. Deduce graphically the stability of the fixed points.

SIR epidemic model

A simple model of the spread of a disease is the *Susceptible, Infected, Recovered* (SIR) model. We consider a population of constant size in the presence of a disease. A member of the population may be in one of three states: susceptible (not yet infected but susceptible to become ill), infected (and therefore contagious), or recovered (not contagious and immune to the disease). The proportions of the population in each of these states are denoted respectively $S(t)$, $I(t)$, and $R(t)$. A very simple model of the dynamics is

$$\begin{cases} \dot{S} = -\beta SI, \\ \dot{I} = +\beta SI - \gamma I, \\ \dot{R} = +\gamma I, \end{cases}$$

where β and γ are the transmission and recovery rates, respectively, and are positive real values.

1. Justify the relation

$$S(t) + I(t) + R(t) = 1.$$

   Deduce that we can limit our study to the two-dimensional system

$$\begin{cases} \dot{S} = -\beta SI, \\ \dot{I} = +\beta SI - \gamma I. \end{cases} \qquad (2.16)$$

2. Plot the region $\mathcal{T}$ of physically acceptable values of $(S(t), I(t))$ in the phase space.
3. Determine the stationary solutions and study their stability. Show that an outbreak from an initially small proportion of infected people $I_0 \ll 1$ is only possible when $\gamma < \beta$. Comment on this result in view of the physical meaning of these parameters.

In the following, we will consider the case $0 < \gamma < \beta$.

4. Determine the explicit solution of system (2.16) for an initial condition of the form $(S_0 = 0, I_0)$. Draw the corresponding trajectory in the phase space.
5. By studying the variation of $S + I$ as a function of time show that for any initial condition (S_0, I_0) in $\mathcal{T}$, the trajectory of the solution $(S(t), I(t))$ remains in $\mathcal{T}$.
6. Draw the nullclines of system (2.16) and the vector field in the region $\mathcal{T}$.
7. Compute the divergence of the vector field and determine the region of the phase space where there is respectively contraction and dilation under the action of the flow.
8. Study the variation of

$$H(S, I) = I + S - \frac{\gamma}{\beta} \ln S$$

   as a function of time and deduce that there is a constant c such that

$$I(t) = c - S(t) + \frac{\gamma}{\beta} \ln(S(t)).$$

   A function such as $H(S, I)$ is called a first integral of the system.
9. Study the function $f(x) = 1 - x + \frac{\gamma}{\beta} \ln(x)$ on the interval $]0, 1]$ and plot its graph in the phase space. What is the relationship between the graph of f and the trajectories of the flow? Draw the shape of all the trajectories in $\mathcal{T}$.

10. Assuming that an epidemic spreads according to the SIR model from a nearly healthy population $S \simeq 1$. According to the model, predict what proportion of people will be infected simultaneously at the peak of the epidemic. At the end of the epidemic, what is the total proportion of the population that has been affected by the disease (and is therefore in recovery)? Draw the shapes of the three variables $S(t)$, $I(t)$ and $R(t)$ as a function of time.

Stationary solutions of the Lorenz model

We introduced in the exercises of Chapter 1 the Lorenz model

$$\dot{X} = \Pr(Y - X),$$
$$\dot{Y} = -XZ + rX - Y,$$
$$\dot{Z} = XY - bZ.$$

We recall that the parameters Pr, r, and b are always positive.
1. What are the fixed points of the system?
2. Compute the Jacobian matrix of the system.
3. Study the stability of the fixed point $(0, 0, 0)$ when $0 < r < 1$ and then when $r > 1$.

Numerical study of a chemical reaction

We have studied in Section 2.1.6.b the stability of the fixed point of the chemical reaction (Strogatz, 2018)

$$\frac{dx}{dt} = a - x - \frac{4xy}{1 + x^2},$$
$$\frac{dy}{dt} = bx\left(1 - \frac{y}{1 + x^2}\right),$$

where a and b are two positive parameters, and x and y are positive variables proportional to the concentration of the reactants.
1. Plot numerically in the phase plane the nullclines and the vector field for $a = 10$ and $b = 4$. Superimpose trajectories starting from different initial conditions.
2. The same question with $a = 10$ and $b = 2$. Superimpose trajectories with initial conditions in the vicinity of the fixed point and observe their behavior.

3 Bifurcations of one-dimensional flows

3.1 Introduction

We have seen in the previous chapters that the most important features of a dynamical system are its invariant sets and attractors. Invariant sets organize trajectories inside the state space of the system and among them, attractors characterize the asymptotic dynamics. They are the backbone of a dynamical system.

In Chapter 1, we considered dynamical systems of the form $\dot{\mathbf{X}} = \mathbf{F}(\mathbf{X})$. Here we take into account the fact that they typically depend on one or several control parameters μ, expressing them as $\dot{\mathbf{X}} = \mathbf{F}(\mathbf{X}, \mu)$. Because attractors and invariant sets have so much influence on the behavior of a system, it is important to understand how they change as the control parameters vary.

Most of the time, invariant sets persist upon variation of a control parameter and are just displaced in the phase space. This is called structural stability. However, we are not interested in how invariant sets are merely displaced, because we are not interested in their exact location, only in their relative organization.

Thus we want to focus on situations where the nature of the invariant sets changes, which we call bifurcations. In bifurcations, invariant sets can appear or disappear, or they can change their stability. The most important bifurcations are those affecting attractors because they modify the asymptotic behavior of the system. The goal is thus to establish a *bifurcation diagram* of a dynamical system, which recapitulates the different states in which it can be depending on parameter values. This is as important as a phase diagram in thermodynamics, which recapitulates the different states in which a substance like water can find itself depending on pressure and temperature.

In this chapter, we will mostly restrict ourselves to the bifurcations that are generically observed in one-dimensional systems when a single control parameter varies. We will see that there are only a few such bifurcation mechanisms, which are universal and do not depend on the details of the system. We will also see how they naturally form the basis of bifurcations occurring in higher-dimensional systems.

Higher-order bifurcations can be observed by tuning simultaneously two or more control parameters. Although they need precise parameter adjustments to occur, they deeply organize the structure of the parameter space around them, as shows the example of the cusp bifurcation (nascent bistability) discussed in Section 3.4.

The Hopf bifurcation, which gives birth to periodic oscillations from a stationary regime, can only occur in two dimensions. It will be studied in Chapter 4 together with bifurcations of these periodic orbits, viewed as bifurcations of fixed points of recurrence systems.

https://doi.org/10.1515/9783110677874-003

3.2 Structural stability in one-dimensional systems

3.2.1 Persistence of an isolated fixed point

The only invariant sets of one-dimensional systems defined by $\dot{x} = f(x, \mu)$ are the fixed points satisfying

$$f(x^*, \mu) = 0 \tag{3.1}$$

When the control parameter μ is varied by $d\mu$, the fixed point is shifted to $x^* + dx^*$ such that the defining equation is still satisfied:

$$f(x^* + dx^*, \mu + d\mu) = 0. \tag{3.2}$$

Expanding this equation to first order, we obtain

$$f(x^*, \mu) + \frac{\partial f}{\partial x}(x^*, \mu)dx^* + \frac{\partial f}{\partial \mu}(x^*, \mu)d\mu = \frac{\partial f}{\partial x}(x^*, \mu)dx^* + \frac{\partial f}{\partial \mu}(x^*, \mu)d\mu = 0,$$

which expresses the implicit function theorem. Hence, in general, we have

$$dx^* = \frac{-\frac{\partial f}{\partial \mu}(x^*, \mu)}{\frac{\partial f}{\partial x}(x^*, \mu)}d\mu. \tag{3.3}$$

This shows that the new fixed point $x^* + dx^*$ is well defined, provided that the denominator $\partial f(x^*, \mu)/\partial x \neq 0$. This is the condition for an isolated fixed point to persist in a neighborhood $[\mu - d\mu, \mu + d\mu]$ of μ. Since this is generally the case, we conclude that, typically, a fixed point that exists for a given parameter value also exists for close parameter values with a small shift in its coordinates. This is called *structural stability*.

3.2.2 Structural instability and bifurcation

Now assume that this is not the case and that

$$\partial f(x^*, \mu)/\partial x = 0. \tag{3.4}$$

Since $x^* = x^*(\mu)$, relation (3.4) represents an equation in μ, so that this singular situation will typically occur for a specific value $\mu = \mu_c$. The Taylor expansion must then be carried out to next order at least, yielding

$$\frac{\partial^2 f}{\partial x^2}(x^*, \mu_c)dx^{*2} + \frac{\partial f}{\partial \mu}(x^*, \mu_c)d\mu = 0.$$

Solving for the displacement dx^* gives

$$dx^{*2} = \frac{-\frac{\partial f}{\partial \mu}(x^*, \mu_c)}{\frac{\partial^2 f}{\partial x^2}(x^*, \mu_c)} d\mu, \tag{3.5}$$

assuming for now that the denominator is not zero. We see that Eq. (3.5) differs radically from (3.3) in that the solutions depend now on the sign of $d\mu$. When the right-hand side of (3.5) is positive, there are two solutions, whereas there are none in the opposite case. For $d\mu = 0$, the two solutions are degenerate and coincide at $dx^* = 0$.

Thus there is an important qualitative change at $\mu = \mu_c$, where the number of fixed points changes. Such a qualitative change is generically called a *bifurcation*. In the present case, two fixed points become degenerate at the bifurcation point and disappear beyond this point (or vice versa, depending on the signs of the right-hand side of (3.5)). This specific bifurcation is called a *saddle-node bifurcation*. In one dimension, the "node" refers to the stable solution, whereas the "saddle" is an unstable solution. In higher dimensions, where the "saddle" terminology is more meaningful, the two solutions are respectively stable and unstable along the direction which separates them.

The saddle-node bifurcation is the simplest and most generic type of bifurcation, as it only depends on the degeneracy condition (3.4), which is common to all bifurcation types. Other bifurcation types typically arise when additional terms in the Taylor expansion of (3.2) become zero, so that the leading terms of this expansion change, giving rise to new equations for the fixed point displacement. Before examining these different bifurcation scenarios in the next section, we now give a geometric interpretation of the bifurcation phenomenon.

3.2.3 Bifurcation and tangencies

Relation (3.4) combined with (3.1) indicates that the bifurcating fixed point at $\mu = \mu_c$ is a tangent intersection of the graph of $f(x)$ with the real axis, since the slope of this graph is zero. Hence the behavior of $f(x)$ around x^* is as depicted in Fig. 3.1, which allows us to understand graphically the structural instability arising at the bifurcation point. It also illustrates how the bifurcation phenomenon can be viewed in terms of the singularities of the intersections between two curves.

In a typical tangent intersection of two curves, each curve lies entirely only on one side of the other curve. Hence any displacement of the curves (such as caused by parameter change) will either remove the intersection or create two transverse intersections. By contrast, transverse intersections, in which curves cross with an angle, are robust: they typically persist under perturbations.

Conversely, it is easy to see that the generic way to add or remove intersections between the graph of $f(x)$ and the real axis is to create a tangency such as shown in Fig. 3.1. The algebraic formulation of the tangency is Eq. (3.4).

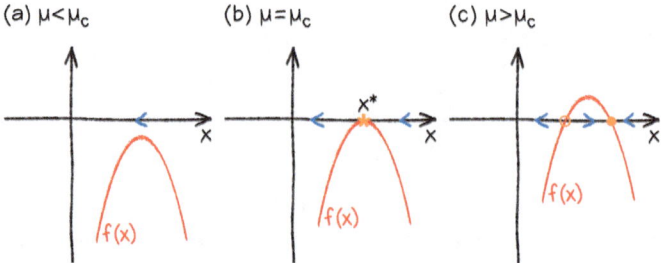

Figure 3.1: Structure of the graph of $f(x)$ in the neighborhood of a saddle-node bifurcation.

Note however that if we also have $\frac{\partial^2 f}{\partial x^2}(x^*, \mu_c) = 0$, which is an additional condition but can forced by symmetry, then two curves can be tangent while crossing each other. This configuration will be at the heart of the pitchfork bifurcation, which we study in Section 3.2.4.c.

3.2.4 Essential bifurcations

To determine the bifurcation scenarios that can occur generically besides the saddle-node one, let us consider more closely the Taylor expansion of $\dot{x} = f(x)$ in a neighborhood of a tangency. For simplicity, and without loss of generality, we assume that the bifurcation is observed for $(x^*, \mu_c) = (0, 0)$, which can always be arranged by redefining x and μ. As discussed above, the defining relations for a bifurcation are

$$f(0, 0) = 0, \quad f_x(0, 0) = 0, \tag{3.6}$$

where we have used the compact notation $f_x = \partial f / \partial x$.

Now let us consider infinitesimally small x and μ, as we will work in a small neighborhood of the bifurcation. The equation $\dot{x} = f(x)$ can then be expanded as

$$\dot{x} = f_\mu \mu + \frac{1}{2} f_{xx} x^2 + f_{x\mu} x\mu + \frac{1}{2} f_{\mu\mu} \mu^2 + \frac{1}{6} f_{xxx} x^3 + \cdots. \tag{3.7}$$

Depending on whether there are terms in (3.7) that are structurally zero, different behaviors are observed. The general strategy is to identify the lowest-order nonzero terms that yield a nontrivial equation. Assuming that the state variable scales as $x \sim \mu^\beta$, the value of β is determined so that at least two terms in (3.7) are of the same order and all other terms are of higher order.

3.2.4.a Saddle-node bifurcation
In the generic case where $f_\mu \neq 0$ and $f_{xx} \neq 0$, the two first terms are of the same order if $\beta = 1/2$, and all other terms are of higher order. Then the dominant part of expan-

sion (3.7) is

$$\dot{x} = f_\mu \mu + \frac{1}{2} f_{xx} x^2.$$

With the rescaling $x = \alpha y$, $\mu = \alpha \mu' / f_\mu$, this leads to the equation

$$\dot{y} = \mu' + \frac{1}{2} \alpha f_{xx} y^2,$$

where α can always be chosen so that it reduces to

$$\dot{y} = \mu' - y^2. \tag{3.8}$$

Equation (3.8) is the simplest dynamical system displaying the saddle-node bifurcation and is called the *normal form* of this bifurcation. Any system displaying this bifurcation can be rewritten in this form in a neighborhood of the bifurcation. From the derivation above this can easily be understood in terms of Taylor expansion, but in fact there is an elaborate mathematical theory showing how to reduce rigorously the system to (3.8) using a series of changes of variable on the state variable and on time. The saddle-node bifurcation will be studied in more detail in Section 3.3.1.

3.2.4.b Transcritical bifurcation

In the saddle-node bifurcation, two fixed points exist on one side of the bifurcation only. In many interesting phenomena, there is a fixed point that exists for all values of the parameter μ, experiencing only a change of stability. Then we can always perform a change of variable so that the fixed point is always at the origin. In this case, as $f(0, \mu) = 0$ for all values of μ, we have that $f_\mu(0,0) = f_{\mu\mu}(0,0) = 0$.

Under these conditions, we obtain the following leading part when $\beta = 1$:

$$\dot{x} = \frac{1}{2} f_{xx} x^2 + f_{x\mu} x \mu = x \left(\frac{1}{2} f_{xx} x + f_{x\mu} \mu \right).$$

Again, this equation can be reduced under a rescaling in x and μ to

$$\dot{y} = \mu y - y^2 = y(\mu - y). \tag{3.9}$$

This is the normal form for a new type of bifurcation, the *transcritical bifurcation*, that will be studied in detail in Section 3.3.2. Fig. 3.2 shows the graphs of the normal form (3.9) before, at and after the bifurcation.

3.2.4.c Pitchfork bifurcation

A key point to be taken into account when deriving a normal form is the existence of symmetries, which will typically restrain the nonzero terms in expansion (3.7). An im-

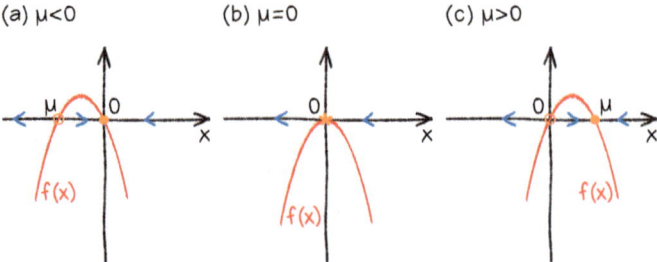

Figure 3.2: Structure of the graph of $f(x)$ in a neighborhood of a transcritical bifurcation.

portant symmetry is the inversion symmetry $x \leftrightarrow -x$, which means that whenever $x(t)$ is a solution of the system, $-x(t)$ is also a solution. That is, if $\dot{x} = f(x)$, then we must have

$$\frac{d}{dt}(-x) = f(-x) = -f(x),$$

which imposes that $f(x)$ is an odd function of x. Thus all terms with an even power of x must be zero, and, in particular, we must have $f_{xx} = 0$. This implies that $f(0) = 0$ for all values of μ, and thus here we also have $f_\mu(0,0) = f_{\mu\mu}(0,0) = 0$. Under these conditions, we obtain the following leading part when $x \sim \mu^{1/2}$:

$$\dot{x} = f_{x\mu}x\mu + \frac{1}{6}f_{xxx}x^3 + \cdots.$$

Again, this expression can be rescaled to

$$\dot{y} = \mu y - y^3 = (\mu - y^2)y, \tag{3.10}$$

which is the normal form of the pitchfork bifurcation that we will study in Section 3.3.3. Fig. 3.3 shows the graphs of the normal form (3.10) before, at and after the bifurcation. In the next Section, we will now study the solutions and bifurcation diagrams associated with the three types of bifurcation that we have identified above.

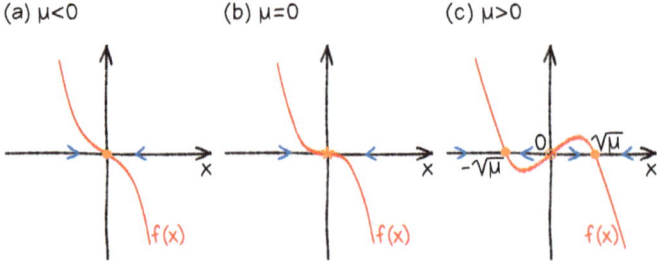

Figure 3.3: Structure of the graph of $f(x)$ in a neighborhood of a supercritical pitchfork bifurcation.

3.3 Generic bifurcations of one-dimensional flows

3.3.1 Saddle-node bifurcation

As we saw in Section 3.2.4.a, the saddle-node bifurcation corresponds to the apparition or annihilation of a pair of fixed points. Its normal form is

$$\dot{x} = \mu - x^2 \quad \text{where } x \in \mathbb{R}.$$

The study of this system can be done graphically from Figure 3.1 or analytically. The possible invariant sets are two fixed points $\pm\sqrt{\mu}$, which exist only when $\mu \geq 0$. As $\frac{\partial f}{\partial x} = -2x$, the fixed point $+\sqrt{\mu}$ is stable, and the other one $-\sqrt{\mu}$ is unstable (see Section 1.3.1). Those fixed points are plotted as functions of the value of the parameter in Figure 3.4. Such a graph is called a *bifurcation diagram*. By convention the set of stable fixed points is drawn with a solid line, and the set of unstable fixed points is drawn with a dashed line.

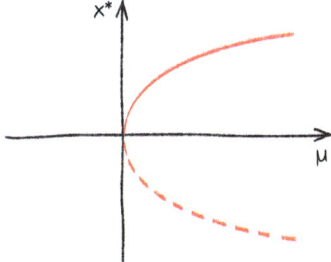

Figure 3.4: Bifurcation diagram of the saddle-node bifurcation. The solid red line is the graph of the function $x^* = +\sqrt{\mu}$, corresponding to the set of stable stationary solutions; the dashed one is the graph of the function $x^* = -\sqrt{\mu}$, which corresponds to the set of unstable stationary solutions.

The two branches of solutions emerge from the bifurcation point $(0, 0)$ at the critical value of the parameter $\mu_c = 0$.

3.3.2 Transcritical bifurcation

This bifurcation corresponds to an exchange of stability between two fixed points. Its normal form is

$$\dot{x} = \mu x - x^2 = x(\mu - x).$$

The fixed points are $x^* = 0$ and $x^* = \mu$. As $\frac{df}{dx} = \mu - 2x$, we find that:

- $\frac{df}{dx}\big|_{x=0} = \mu$, the fixed point $x = 0$ is stable for $\mu < 0$ and unstable for $\mu > 0$;
- $\frac{df}{dx}\big|_{x=\mu} = -\mu$, the fixed point $x = \mu$ is unstable for $\mu < 0$ and stable for $\mu > 0$.

The bifurcation diagram of the transcritical bifurcation is shown in Fig. 3.5.

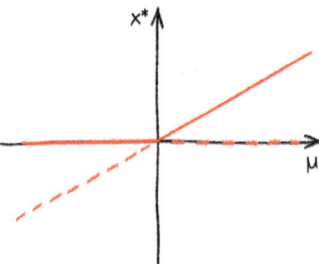

Figure 3.5: Bifurcation diagram of the transcritical bifurcation.

3.3.3 Pitchfork bifurcation

3.3.3.a Supercritical bifurcation

As discussed in Section 3.2.4.c, when a system has an inversion symmetry $x \leftrightarrow -x$, we obtain the normal form

$$\dot{x} = \mu x - x^3 = x(\mu - x^2).$$

Depending on the sign of μ, there are 1 or 3 fixed points: the fixed point $x^* = 0$ exists for all values of μ, whereas the two symmetric fixed points $\pm\sqrt{\mu}$ exist only for $\mu > 0$.

The stability analysis gives:

- $\frac{df}{dx}\big|_{x=0} = \mu$, the fixed point 0 is stable when $\mu < 0$ and unstable when $\mu > 0$;
- $\frac{df}{dx}\big|_{x=\pm\sqrt{\mu}} = -2\mu < 0$, the two symmetric fixed points $\pm\sqrt{\mu}$ are stable when they exist.

This leads to the bifurcation diagram shown in Fig. 3.6.

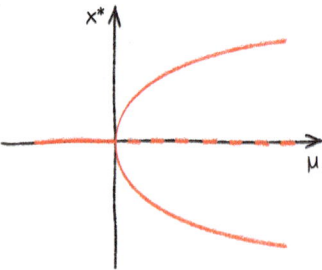

Figure 3.6: Bifurcation diagram of the supercritical pitchfork bifurcation.

When the system crosses the bifurcation coming from $\mu < 0$, it has to choose one of the two stable branches of the pitchfork. This phenomenon is called a *symmetry breaking* as the emerging solution has lost the symmetry $x \leftrightarrow -x$. Note that the set of all the solutions is still symmetric.

3.3.3.b Subcritical bifurcation

For each normal form, we may ask what is the effect of changing the sign of the non-linearity. In the case of the saddle-node and transcritical bifurcations, which have a quadratic nonlinearity, the new expression can be brought back to the original one by changing the variable and redefining the parameter.

In the case of the pitchfork bifurcation, this leads to the nonequivalent expression $\dot{x} = \mu x + x^3 = x(\mu + x^2)$. In the supercritical case we have discussed above, the nonlinear term $-x^3$ saturates the linear divergence for $\mu > 0$, leading to two new stable solutions that coexist with the unstable fixed point at $x^* = 0$. In the subcritical case, there is no stable fixed point to serve as an attractor when the origin becomes unstable, as can be seen in Fig. 3.7, and the system diverges to infinity. Here the term *subcritical* refers to the fact that the bifurcating solutions exist before the fixed point destabilizes at the (critical) bifurcation point.

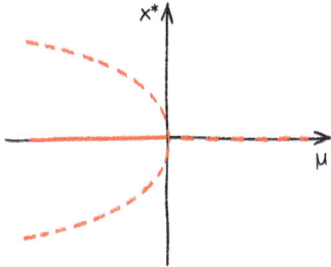

Figure 3.7: Bifurcation diagram of the normal form $\dot{x} = \mu x + x^3$.

Consequently, to obtain stable solutions for $\mu > 0$, we have to introduce higher-order terms in the normal form to saturate the instability. The simplest solution preserving the $x \to -x$ symmetry is

$$\dot{x} = \mu x + x^3 - x^5. \tag{3.11}$$

To obtain the corresponding bifurcation diagram, we have to study the fixed points of Eq. (3.11) and their stability.

The solutions of $\mu x + x^3 - x^5 = 0$ are either $x^* = 0$ or the solutions of the polynomial $\mu + x^2 - x^4 = 0$. Letting $y = x^2$, we search the solutions of $y^2 - y - \mu = 0$, which are $y_\pm = \frac{1 \pm \sqrt{1+4\mu}}{2}$ for $\mu \geq -1/4$. As $y = x^2 \geq 0$, these solutions are meaningful only when $y_\pm$ are positive.

- $y_+ = \frac{1 + \sqrt{1+4\mu}}{2}$ is always positive in its existence domain $\mu \geq -1/4$;

- $y_- = \frac{1-\sqrt{1+4\mu}}{2}$ is positive only when $1 \geq \sqrt{1+4\mu}$, i. e., when $-1/4 \leq \mu \leq 0$.

To summarize (Fig 3.8):
- when $\mu < -1/4$, there is only one stationary solution $x_0^* = 0$, and as $\frac{df}{dx}\big|_0 = \mu < 0$, it is stable;
- when $-1/4 \leq \mu \leq 0$, there are five fixed points. The solution $x_0^* = 0$ is still stable. The four other fixed points are given by $\pm\sqrt{y_+}$ and $\pm\sqrt{y_-}$. We can check by a linear analysis that the two solutions $x_{1\pm}^* = \pm\sqrt{\frac{1-\sqrt{1+4\mu}}{2}}$ are unstable, whereas the solutions $x_{2\pm}^* = \pm\sqrt{\frac{1+\sqrt{1+4\mu}}{2}}$ are stable. This was expected since we know that the stability of fixed points alternates along the real axis;
- when $\mu \geq 0$, only three fixed points remain because the solutions $\pm\sqrt{y_-}$ no longer exist. The fixed point $x_0^* = 0$ is now unstable as $\mu > 0$, whereas the solutions $x_{2\pm}^* = \pm\sqrt{\frac{1+\sqrt{1+4\mu}}{2}}$ are still stable.

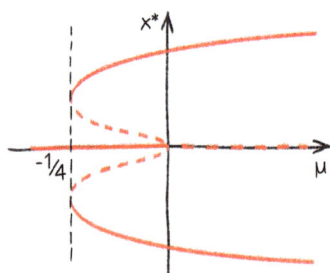

Figure 3.8: Bifurcation diagram of the subcritical pitchfork bifurcation.

As a matter of fact, the two solutions $x_{2\pm}^*$ are stable on their whole domain of existence, whereas $x_{1\pm}^*$ are always unstable. The bifurcation where the stable $x_{2\pm}^*$ and the unstable $x_{1\pm}^*$ appear together is familiar to us: it is a saddle-node bifurcation. The symmetric scenario is observed for x_{2-}^* and x_{1-}^* on the other side of the x-axis. Note how all solutions are connected to each other, with in particular the unstable solutions $x_{1\pm}^*$ connecting the pitchfork bifurcation at the origin with the saddle-node bifurcation where the stable solutions $x_{2\pm}^*$ appear.

In the interval $[-1/4; 0]$, several stable solutions coexist: there is *bistability* between the solutions. The choice between the zero solution and one or the other of the symmetric stable branches depends on the history of the system. If the parameter μ is tuned from a negative value smaller than $-1/4$, then the system will follow the stable zero branch as long as $\mu < 0$. When μ crosses 0, the system will shift on one of the nonzero branches. If μ now decreases, then the system will stay on the nonzero branch until $\mu = -1/4$, where the system will switch on the $x = 0$ branch. Such a behavior is called a *hysteresis cycle*.

3.4 An example of higher-order bifurcation

Until now, we have restricted ourselves to the generic bifurcations observed when a single parameter is varied. Here we will study an example of a higher-order bifurcation, which occurs generically only when two parameters are varied.

We consider here the normal form of pitchfork bifurcation, where we break the inversion symmetry by introducing a nonzero term v:

$$\dot{x} = \mu x - x^3 + v. \tag{3.12}$$

We will now study how this one-dimensional bifurcation is modified when we simultaneously vary μ and v.

3.4.1 Fixed points

To study graphically the fixed points of Eq. (3.12), we consider the intersections of the curve $y = g(x) = x^3 - \mu x$ with the horizontal line $y = v$. An important point is how the shape of $g(x)$ depends on the sign of μ.

When $\mu < 0$, $g(x)$ is a strictly increasing function, and only one fixed point x^* exists for all values of v (Fig. 3.9a).

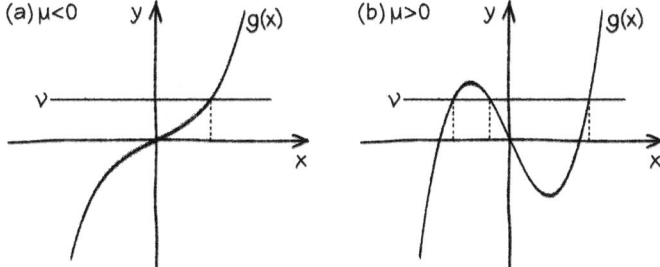

Figure 3.9: Finding the fixed points of $\dot{x} = \mu x - x^3 + v$ by looking at the intersections of the lines $y = v$ with the curve $y = g(x) = x^3 - \mu x$ in the cases (a) $\mu < 0$ and (b) $\mu > 0$.

When $\mu > 0$, the derivative of $g(x)$ is negative in the interval $[-\sqrt{\frac{\mu}{3}}, +\sqrt{\frac{\mu}{3}}]$ (see Fig. 3.9b). Depending on the value of v, we can find either one or three fixed points. A straightforward calculation indicates that we have three fixed points when $v \in [-\frac{2\mu}{3}\sqrt{\frac{\mu}{3}}, +\frac{2\mu}{3}\sqrt{\frac{\mu}{3}}]$ (Fig. 3.9b), whereas there is only one fixed point when v is outside this interval.

Gathering those results, we can then draw a *stability diagram* in the parameter space (μ, v), as shown in Figure 3.10. In the hatched area, there are three fixed points, and only one fixed point outside. This area is delimited by the curves $v = \frac{2\mu}{3}\sqrt{\frac{\mu}{3}}$ and $v = -\frac{2\mu}{3}\sqrt{\frac{\mu}{3}}$. Their junction point in $(0, 0)$ is a singular point called a *cusp*.

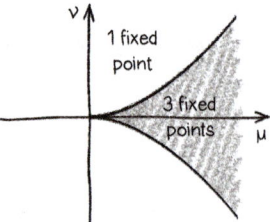

Figure 3.10: Stability diagram of the imperfect pitchfork bifurcation.

3.4.2 Stability of the fixed points

To determine the stability of the fixed points, we recall that $f(x) = \mu x - x^3 + v$, and thus $df/dx = -dg/dx$, so that the stability of fixed points can be deduced from the sign of the derivative of g. In the case $\mu < 0$ the only existing fixed point is stable. When $\mu > 0$, the fixed points obtained in the increasing part of g (the rightmost and leftmost ones) are stable, whereas the one obtained in the decreasing part (middle fixed point) is unstable.

This allows us to draw different types of bifurcation diagrams: a bifurcation diagram for fixed μ and varying v, and another one for fixed v and varying μ.

At fixed $\mu < 0$, no bifurcation occurs when v is varied (Fig. 3.9a). At fixed $\mu > 0$, saddle-node bifurcations occur when the line $y = v$ is tangent with the curve $y = g(x)$ in Fig. 3.9b, i. e., for $v = \pm\frac{2\mu}{3}\sqrt{\frac{\mu}{3}}$. The corresponding bifurcation diagrams are shown in Fig. 3.11.

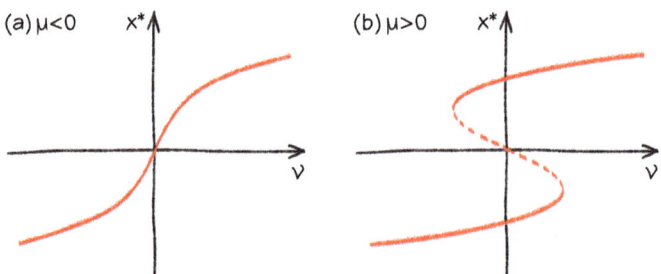

Figure 3.11: Bifurcation diagram at fixed μ; (a) $\mu < 0$, (b) $\mu > 0$.

At fixed v the bifurcation diagram can be drawn by gradually deforming the curve $y = g(x)$ while keeping the line $y = v$ constant. For $v > 0$, we hence obtain Fig. 3.12.

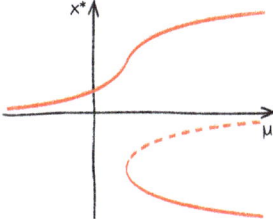

Figure 3.12: Bifurcation diagram of the imperfect pitchfork bifurcation (case $v > 0$).

This bifurcation is called an *imperfect pitchfork bifurcation*. If we compare this diagram to that of Fig. 3.6, the set of solutions has been divided in two parts: a positive fixed point, which always exists and is stable, and a pair of negative fixed points that emerge from a saddle-node bifurcation. The perfect symmetry of the pitchfork bifurcation has disappeared.

In fact, the imperfect bifurcation scenario corresponds to what is observed in a real experiment. An archetypal example of a pitchfork bifurcation is the buckling of a beam under load. Consider a real vertical beam loaded at its top with a weight, and suppose that it is not perfectly symmetric (see Fig. 3.13). If the weight is small, then the beam will be (almost) vertical and straight. If the weight exceeds a critical value, then the beam does not buckle randomly on one side or the other at each new experiment. On the contrary, it tends to buckle always on the same side, which is determined by the imperfections that break the symmetry of the system (vertical misalignment, beam imperfection, etc.). Such a behavior corresponds to always following the same branch in the bifurcation diagram when varying the force at the top of the beam, i. e., the upper branch of Fig. 3.12. To observe buckling on the other side, the system must be subjected to a large perturbation, which will cause it to jump abruptly to the lower stable branch.

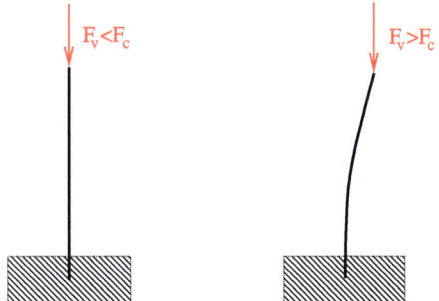

Figure 3.13: Buckling of a strut. A vertical thin beam is submitted to a compressive force F. If F is small enough, then the beam stays straight and sustains the force. Above a critical value F_c, the beam buckles: it bends elastically on one side or the other to minimize its internal energy. In theory, buckling corresponds to a symmetry breaking: the beam can bend with the same probability on one or the other side. In a real experiment, we observe that it bends always on the same side due to imperfections.

3.4.3 Cusp catastrophe

A global representation of the stationary solutions as a function of the two parameters (μ, v) is displayed in Figure 3.14. We can see that the surface of solutions is regular in the three-dimensional space but that its projection on the (μ, v)-plane displays singularities. This is an example of a higher-order bifurcation that results from the interaction of two elementary bifurcations. The area of the parameter plane where three solutions coexist is delimited by two *fold lines* associated with the saddle-node bifurcations: in a saddle-node bifurcation, two solutions are folded over each other. When the two fold singularities collide, this gives rise to a cusp singularity (or a cusp catastrophe). Note that two fold lines are not transverse at a cusp catastrophe but meet tangentially. This is an essential ingredient of the higher-order singularity.

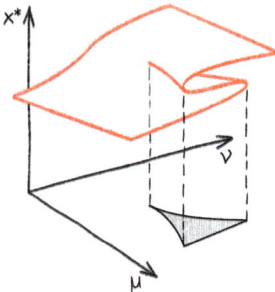

Figure 3.14: The set of solutions of Eq. (3.12) is shown as functions of the two parameters (μ, v). The projection of the surface of solutions on the two-dimensional parameter plane is also shown. At the intersections of the two fold lines, it displays a cusp singularity, also known as a cusp catastrophe.

Singularities in functions or dynamical systems follow a hierarchical organization, which is the subject of *singularity theory*, also known as *catastrophe theory* (Arnold et al., 2012; Gilmore, 1981).

3.5 One-dimensional bifurcations in higher dimensions

3.5.1 Local analysis

The bifurcations of a system $\dot{x} = f(x, \mu)$ that we have studied in this chapter are associated with the condition $\partial f(x, \mu)/\partial x = 0$ for $(x, \mu) = (x^*, \mu_c)$, stating that the linear part of the system vanishes. Recalling our study of linear stability in Chapter 2, this is associated with the only eigenvalue of the one-dimensional system crossing zero. This condition has important consequences not only on the stability of solutions, but also on their existence due to the resulting structural instability.

In higher-dimensional systems, elementary bifurcations of a fixed point are associated with the vanishing of the real part of a single eigenvalue or of a pair of complex conjugate eigenvalues. Excluding the latter case, which will be treated in Section 4.1.1, we will assume here that a single eigenvalue changes sign as a control parameter is varied. As we will illustrate, qualitative changes are then restricted to the direction of the corresponding eigenvector, along which the Jacobian is singular. This ensures that the one-dimensional analysis developed in this chapter applies in any phase space dimension.

For simplicity, we consider a two-dimensional system with a single real eigenvalue crossing zero at the bifurcation. If the Jacobian is diagonalizable, then we can find a system of coordinates such that in a neighborhood of the bifurcation and of the bifurcating fixed points, the system can be written in the following form:

$$\dot{x} = f(x, \mu), \tag{3.13a}$$

$$\dot{y} = \lambda y, \tag{3.13b}$$

where the x-axis corresponds to the direction along which an eigenvalue crosses 0, whereas the y-direction corresponds to an eigenvalue λ that does not change sign when μ varies (here we assume that $\lambda < 0$ for simplicity). In Eqs. (3.13), $f(x, \mu)$ is one of the normal forms that we have identified in Section 3.2.4 and studied in Section 3.3, and which thus satisfies $f(0, 0) = 0$ and $\frac{\partial f}{\partial x}(0, 0) = 0$. Indeed, once the dynamics along the x- and y-directions have been decoupled, the analysis can proceed as described in the previous sections of the current chapter. Thus the different bifurcation scenarios studied previously, describing how solutions change their stability or appear/disappear, remain relevant.

The case of the saddle-node bifurcation is illustrated in Fig. 3.15, where we show the configuration of the vector field around the fixed points before, during, and after the bifurcation.

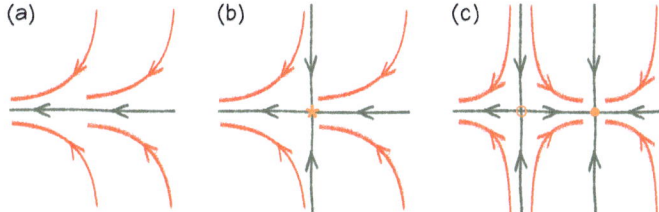

Figure 3.15: Saddle-node bifurcation in two dimensions. Along the vertical direction, corresponding to the y-axis in Eq. 3.13, $\lambda < 0$, whereas along the horizontal direction, corresponding to the x-axis in Eq. (3.13), a saddle-node bifurcation takes place.

3.5.2 Example of a two-dimensional saddle-node bifurcation

It is also interesting to consider the global unfolding of a saddle-node bifurcation in the two-dimensional phase plane using nullclines. For this, we consider a genetic circuit based on a gene activated by the protein it produces. A model of this circuit can be written as

$$\dot{x} = \frac{y^2}{1+y^2} - x, \tag{3.14a}$$

$$\dot{y} = kx - dy, \tag{3.14b}$$

where x and y represent the concentrations of the mRNA and protein molecules produced from the self-activating gene, and where the two variables and time have been suitably normalized to reduce the number of parameters ($k, d > 0$).

Using algebraic manipulations, we can show that the positions of the fixed points depend only on the ratio $\rho = k/d$ and that in addition to the obvious fixed point $(x, y) = (0, 0)$, there are two other fixed points

$$y = \frac{\rho \pm \sqrt{\rho^2 - 4}}{2} \quad \text{and} \quad x = \rho y,$$

which exist when $\rho \geq 2$. This indicates that a saddle-node bifurcation occurs for $\rho = 2$. However, it is rarely the case that fixed points can be solved algebraically in such a simple way.

Graphically, the fixed points are found at the intersection of the two curves $x = f_1(y) = \frac{y^2}{1+y^2}$ and $x = f_2(y) = \frac{1}{\rho}y$. One way to find the bifurcation is to search for the tangency of the two curves by requesting that the two curves coincide and have identical slope:

$$f_1(y) = f_2(y), \quad \frac{df_1(y)}{dy} = \frac{df_2(y)}{dy}.$$

Thus we can answer the question by eliminating y between the two equations:

$$\frac{y^2}{1+y^2} = \frac{1}{\rho}y, \quad \frac{2y}{(1+y^2)^2} = \frac{1}{\rho},$$

which leads to $y = 1$, $\rho = 2$. Similar equations are obtained if we express the fact that the determinant of the Jacobian matrix should be zero at the fixed point corresponding to the zero eigenvalue.

If we are not interested in the exact values of the fixed points, nor in the exact values of the bifurcating parameters but only in understanding the bifurcation scenario, then it is sufficient to plot the two curves $f_1(y)$ and $f_2(y)$ for different values of ρ as shown in Fig. 3.16.

We now show that the stability of the fixed points can be deduced from the graphs of the two functions $f_1(y)$ and $f_2(y)$. Indeed, the Jacobian matrix of Eqs. (3.14) can be writ-

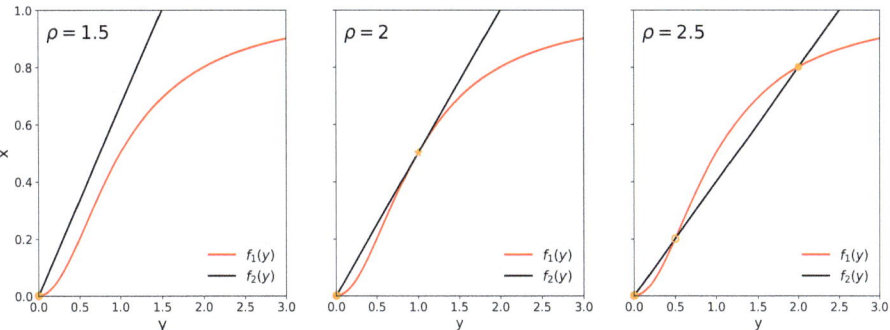

Figure 3.16: Graphical determination of the fixed points of Eqs. (3.14) for different values of ρ.

ten as

$$\mathcal{L} = \begin{pmatrix} -1 & f_1'(y) \\ k & -kf_2'(y) \end{pmatrix},$$

so that its trace and determinant are respectively $T = -1 - kf_2'(y)$ and $\Delta = k(f_2'(y) - f_1'(y))$. Since we see in Fig. 3.16 that f_1 and f_2 are both increasing functions, we can conclude that $T < 0$ for all y. As $\Delta > 0$ (resp., < 0) when $f_2' > f_1'$ (resp., $f_2' < f_1'$), the fixed point for which the slope of f_1 is larger than that of f_2 is unstable, whereas those for which $f_2' > f_1'$ are stable. Another way to find the stability is to draw the vector field using the fact that f_1 and f_2 are the nullclines of the system.

Interestingly, we see that as the point $(0, 0)$ is always stable, a finite amount of protein is needed for gene to self-activate, bringing the system beyond the unstable fixed point, which separates the basins of attraction of the two stable fixed points.

3.6 Conclusions

In one-dimensional systems the only invariant sets are fixed points. In this chapter, we have studied qualitative changes of fixed points, termed *bifurcations*, in which they can appear or disappear, or change their stability. Bifurcations are typically associated with a degeneracy of fixed points or, more generally, of invariant sets.

We have identified three main types of bifurcations: the saddle-node, transcritical, and pitchfork bifurcations, which can generally occur when a single parameter is varied. The first one is generic, and the other two occur when additional constraints are imposed. These bifurcations are also encountered in higher-dimensional systems when one real eigenvalue crosses zero, which make them ubiquitous in nonlinear systems.

When two parameters can be varied, it becomes possible to make elementary bifurcations coincide, in which case a higher-order bifurcation is observed.

Exercises

Study of a mechanical system

Consider a bead of mass m placed in a circular guide that rotates around a vertical axis at a constant angular velocity ω (Fig. 3.17).

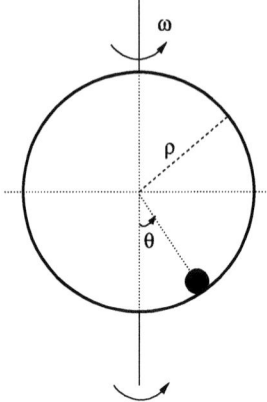

Figure 3.17: Bead in a circular guide.

The guide forms a ring of radius ρ on which the bead is constrained to stay. The bead diameter is negligible compared to ρ. The angle marking the position of the bead with respect to the vertical axis is θ, which can take values between $-\pi$ and π. We consider that a viscous force $F_v = -b\frac{d\theta}{dt}$ dampens the dynamics, and we denote by g the gravitational acceleration.

1. Using the conservation of momentum, show that the dynamics of the bead is given by the following differential equation:

$$m\rho\frac{d^2\theta}{dt^2} = -b\frac{d\theta}{dt} - mg\sin\theta + m\omega^2\rho\sin\theta\cos\theta. \tag{3.15}$$

2. Nondimensionalize the system by rescaling the time as $\tau = \omega t$. We will denote in the following $\frac{d\theta}{d\tau} = \dot{\theta}$ and we will pose $\gamma = g/(\rho\omega^2)$ and $\beta = b/(m\rho\omega)$. What sign can these parameters take?

3. Rewrite Eq. (3.15) as a first-order dynamical system.

4. Determine the fixed points of the system. For which values of γ and β do those fixed points exist? Represent on schematics the experimental configurations corresponding to the fixed points. Discuss the symmetries of the system.

5. Compute the Jacobian matrix.

6. Study the stability of each of the fixed points. Give the type of each fixed point (node, focus, saddle point) depending on the values of γ and β. What is the effect of damping on the nature of the fixed points?
7. What kind of bifurcation occurs in the system?
8. In practice, in an experiment the tunable parameter is ω, whereas the other parameters ρ, g, b, g are constant. Draw the bifurcation diagram of the stationary solutions θ^* as a function of ω.

A model of class B laser

The following system models a two-level laser:

$$\begin{cases} \dot{D} = -D(I + 1) + a, \\ \dot{I} = kI(D - 1), \end{cases}$$

where D is the population inversion, which describes the state of the amplification medium, $I \geq 0$ is the laser intensity, the parameter k is constant and depends on the properties of the gain medium and of the losses of the optical cavity, and a is the pumping rate, the tunable parameter in an experiment.

1. Determine the fixed points of the system and study their stability as a function of the parameter a. Which solutions correspond to the laser off or on?
2. Draw the bifurcation diagrams for the variables I and D.

A model of class C laser

A more general model of laser is given by the Maxwell–Bloch equations

$$\dot{E} = -\kappa(E + P),$$
$$\dot{P} = -\gamma_1(P + ED),$$
$$\dot{D} = -\gamma_2(D - a - EP),$$

where E is the electric field, P is the average polarization of the atoms in the amplification medium, and D is the inversion of population in this medium. The parameters $\kappa, \gamma_1, \gamma_2$, and a are all strictly positive, κ is related to the optical cavity losses, γ_1 is the polarization decay rate, γ_2 is the population inversion decay rate, and a is the pumping rate. This last parameter is the only one that can be easily tuned experimentally.

1. Determine the fixed points of the system. Which solutions correspond to the laser off or on?
2. Study the stability of the fixed point when the laser is switched off as a function of the parameter a.

3. Draw the bifurcation diagrams for the different variables and identify the type of bifurcations observed.
4. Using the change of variables

$$t' = \gamma_1 t, \quad X = -\sqrt{\frac{\gamma_2}{\gamma_1}} E, \quad Y = \sqrt{\frac{\gamma_2}{\gamma_1}} P, \quad \text{and} \quad Z = a - D,$$

show that this laser model can be rewritten as the Lorenz model.

Bullard dynamo

This exercise is based on the article "Chaotic motors", C. Laroche et al., *Am. J. Phys.* **80**, 113 (2012). A follow-up to this exercise can be found in the Chapter 4.

This problem deals with the question of the direct conversion of mechanical work in electricity based on self-induction in the absence of an imposed exterior magnetic field or current.

Consider the schematic of a dynamo in Fig. 3.18. A conducting disk is set in rotation at a rotation rate Ω by a driving torque Γ. If a magnetic field **B** was imposed, a current I would be induced in the wire loop around the axis of rotation. Conversely, if a current circulates in the coil, then a magnetic field **B** is induced. Now the question is: can a current I (and a magnetic field **B**) emerge spontaneously by the mere rotation of the conducting disk in the configuration of Fig. 3.18?

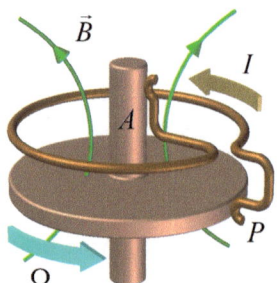

Figure 3.18: Reprinted from (Laroche et al., 2012) with the permission of AIP Publishing. Sketch of the Bullard dynamo.

The equation describing the evolution of the current circulating in the coil is

$$L\frac{dI}{dt} + RI = M\Omega I, \tag{3.16}$$

where R and L are respectively the resistance and inductance of the circuit, and M is the mutual inductance between the coil and the conducting disk.

1. What is the criterium on the rotation rate Ω for a current to spontaneously emerge in the coil?

The disk driven by an imposed torque Γ decelerates because of both mechanical friction and induced currents in the disk. The equation governing the rotation rate Ω can then be written as follows:

$$J\frac{d\Omega}{dt} = \Gamma - \lambda\Omega - MI^2, \qquad (3.17)$$

where J is the moment of inertia of the disk, and λ is the friction coefficient that accounts for mechanical friction losses.

2. Determine the fixed points of the system constituted of the coupled Eqs. (3.16) and (3.17) and study their stability.
3. Draw the bifurcation diagrams for I and Ω when the parameter Γ varies.

Euler strut

In this problem, we detail an archetypal example of the pitchfork bifurcation, the buckling of an elastic beam (see Fig. 3.13), and we will follow the approach of (Barber and Loudon, 1989). The treatment of the bifurcation will be different from the method we used in the previous exercises. We will study the static equilibrium of a vertical beam submitted to a compressive force and show that above the critical value of this force, the strut buckles for an infinitesimal lateral perturbation. This example will allow us to study several aspects of bifurcations as the critical slowing down of vibrations in the vicinity of a bifurcation as well as the analogy between the supercritical pitchfork bifurcation and second-order phase transition. Finally, this system provides a good example of imperfect bifurcation and thus of cusp catastrophe.

We consider a vertical slender beam clamped at its base that can support its own weight (see Fig. 3.13). Its transverse dimensions are very small compared to its length L. A compressive vertical force F_v is applied at the top of the beam. If the force is small enough, then the beam is laterally stable: it holds its vertical position and returns to this equilibrium position if we slightly shift the top of the beam and then release it. For a large enough compressive force, the column buckles, i. e., it takes a new equilibrium position.

Buckling of a column: static approach
In this first part, we study the lateral force F_l needed to displace the top of the beam of δ while applying a compressive vertical force F_v. We will neglect the weight of the beam compared to the compressive forces applied. We also assume that the deflection δ of the top of the rod is small ($\delta \ll L$).

Let us consider a slightly deformed beam (see Fig. 3.19). Each segment of the strut is submitted to a torque and reacts by producing an internal bending moment M that

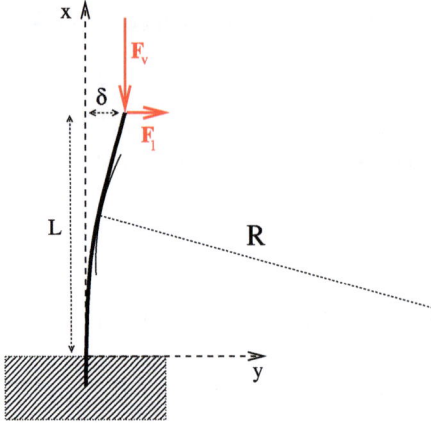

Figure 3.19: Configuration studied: the beam is slightly deformed under the action of both compressive and lateral forces. The new shape of the beam is given by the curve $y(x)$.

balances the torque due to the external forces. This moment is $M = EI/R$ with $1/R$ the local curvature of the beam, E the Young's modulus of the material, and I a geometrical factor: the second moment of area with respect to the y-axis. We note $y(x)$ the displacement of the beam from its undistorted position. For small enough strain ($|dy/dx| \ll 1$) we have $1/R = d^2y/dx^2$. Consequently (Barber and Loudon, 1989),

$$M = EI\frac{d^2y}{dx^2}.$$
(3.18)

1. Consider a section of the beam at position (x, y). Show that the torque produced by the applied force at that point is

$$M = F_v(\delta - y) + F_l(L - x).$$
(3.19)

2. Deduce from Eqs. (3.18) and (3.19) that the shape of the deformed beam is given by

$$y = C_1 \cos kx + C_2 \sin kx + \frac{\delta F_v + (L - x)F_l}{F_v},$$

where $k = \sqrt{\frac{F_v}{EI}}$. C_1 and C_2 are constants that will be determined in the following.

3. The clamped boundary condition (no displacement and no rotation) at the bottom of the strut is:

$$y(0) = 0 \quad \text{and} \quad \frac{dy}{dx}(0) = 0.$$

Find the values of C_1 and C_2.

We have now the expression of the deflection $y(x)$ of the beam over its entire height. However, it depends on the value of δ, which is not independent of F_v and F_l.

4. Show that

$$\delta = \frac{(\tan kL - kL)F_l}{EIk^3}. \tag{3.20}$$

5. Plot δ/F_l as a function of $kL \in [0, \frac{\pi}{2}[$.
6. Discuss the behavior of the beam when k is close to $k_c = \frac{\pi}{2L}$. In particular, discuss the physical meaning of the ratio F_l/δ and what it means for this ratio to cancel out. Give the expression of the critical load for which the strut buckles and justify that the bifurcation is a pitchfork bifurcation.

Beam's vibration: critical slowing down

In this section, we study the frequencies of the oscillations around the equilibrium vertical position before buckling, i. e., when $F_v \lesssim \frac{\pi^2 EI}{4L^2}$. The geometry is the same as before: a beam is positioned vertically with its lower end firmly clamped. A mass m at the origin of a compressive force $F_v = mg$ is fixed to the top of the beam.

1. Using the expressions established in the previous part to deduce the restoring force produced by a deflection δ, show that the equation of motion governing the mass m is

$$m\frac{\mathrm{d}^2\delta}{\mathrm{d}t^2} = -\frac{mgk}{\tan kL - kL}\delta. \tag{3.21}$$

2. Show that the frequency of oscillation is given by

$$\omega = \sqrt{\frac{g}{L}\frac{kL}{\tan kL - kL}}. \tag{3.22}$$

3. Show that we can define a critical mass m_c such that close to the bifurcation,

$$\omega \simeq \sqrt{g(k_c - k)k_c L}$$
$$\simeq \frac{\pi}{2\sqrt{2}}\sqrt{\frac{g}{L}}\left(1 - \frac{m}{m_c}\right).$$

Explain why we can speak of a *critical slowing down at the bifurcation*.

Analogy between the pitchfork bifurcation and a second-order transition

Finding the expression of the deflection δ of the strut after the bifurcation is difficult. Instead, we consider a simplified system that contains the essence of the physics at play (Barber and Loudon, 1989). Instead of considering an elastic beam that undergoes buckling, we will consider a rigid rod attached to the floor by a small elastic strip that plays the role of a spring and that resists deflection of the rod with a restoring torque $\kappa\theta$ (see Fig. 3.20). In the following, the angle $\theta = \delta/L$ will be considered as small.

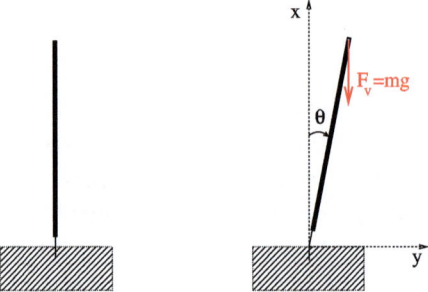

Figure 3.20: Simplified model: instead of an elastic beam, we consider a rigid one fixed to the ground by a small elastic strip.

1. Show that the equation of motion of a mass m fixed at the top of the rod is

$$mL\ddot{\theta} = -\kappa\theta + mgL\sin\theta \tag{3.23}$$

$$\simeq (mgL - \kappa)\theta - mgL\frac{\theta^3}{6}. \tag{3.24}$$

2. Find the equilibrium solutions of this system and study their stability.
3. What is the critical value m_c of m at which the bifurcation takes place?
4. What must be the value of the spring constant κ to recover the critical load obtained in the previous sections?

In the following, we take the spring constant found in the previous question.
5. Show that the total potential energy of the system is

$$U(\theta) = \frac{1}{2}\kappa\theta^2 + mgL(\cos\theta - 1) \tag{3.25}$$

$$\simeq \left(1 - \frac{m}{m_c}\right)m_c gL\frac{\theta^2}{2} + mgL\frac{\theta^4}{24}. \tag{3.26}$$

6. Draw the potential $U(\theta)$ for $m < m_c$, $m = m_c$, and $m > m_c$ and discuss the equilibrium positions of the system and their stability.
7. Compare this system to a system undergoing a second-order transition as, for example, a ferromagnetic material that acquires a spontaneous magnetization below a critical temperature.

Catastrophe theory

Here we still consider the simplified system of a rigid rod attached to the ground by an elastic strip, but we now consider the situation where a small lateral force $F_l \ll mg$ is applied at the top of the rod in addition to the weight of mass mg.

1. Show that the equation of motion of the mass is then given by

$$mL\ddot{\theta} = -\kappa\theta + mgL\sin\theta + LF_l\cos\theta \tag{3.27}$$

$$\simeq F_l L + (mgL - \kappa)\theta - mgL\frac{\theta^3}{6}. \tag{3.28}$$

2. Show that the fixed points of the system can be obtained by studying graphically the intersections of an horizontal line $y = F_l L$ with the function

$$g(\theta) = mgL\left[\frac{\theta^3}{6} - \left(1 - \frac{m_c}{m}\right)\theta\right].$$

3. Show that for a given value of the load mg, the regions where there exist one root and three roots are separated by a critical value of the applied force given by

$$F_l^{(c)} = \pm mg\sqrt{\frac{8}{9}}\left(1 - \frac{m_c}{m}\right)^{3/2}.$$

Draw the stability diagram in the parameter space (m, F_l) (see Section 3.4).

4. Write Eq. (3.28) as a system of first-order equations and show that the Jacobian matrix of this system is given by

$$\mathcal{L} = \begin{pmatrix} 0 & 1 \\ -g'(\theta) & 0 \end{pmatrix}.$$

5. Discuss the stability of the fixed points as a function of the slope of $g(\theta)$. What physical ingredient is missing in our model? Justify that in the following we will consider that the fixed points for which $g'(\theta^*) > 0$ are stable.

6. Consider a fixed value $m > m_c$. Draw the general shape of the bifurcation diagram θ^* as a function of F_l. Explain how hysteresis can be observed in this system by varying the lateral force exerted at the top of the strut.

7. Using Section 3.4, draw the set of equilibrium solutions as a function of the parameters mg and F_l.

4 Oscillations

4.1 Oscillations and limit cycles

4.1.1 Birth of oscillations through the Hopf bifurcation

In Chapter 3, we made the point that oscillations cannot arise in 1D systems, in which the motion can only be monotonous. We also stressed in Section 3.5 that 1D bifurcations of fixed points can appear in higher dimensions and that they typically are associated with the real part of a single eigenvalue crossing zero.

Here we study how oscillations can emerge from the destabilization of a fixed point through a bifurcation that is essentially 2D, involving simultaneously two eigenvalues. If the scenario is to be generic by varying a single parameter, these eigenvalues cannot be independent: the natural situation is that of two complex conjugate eigenvalues sharing the same real part, which becomes zero at the bifurcation. In Section 2.2.3, we already discussed this situation, which was depicted in Fig. 2.7(c). Therefore the two eigenvalues must be of the form $\lambda_{1,2} = \pm i\omega$ at the bifurcation, with the corresponding linear system most simply expressed in the complex plane as

$$\dot{z} = (\mu + i\omega)z \qquad (4.1)$$

with $\mu = 0$ at the bifurcation. In this case the solution $z(t) = e^{i\omega t}$ clearly points to an oscillatory behavior. The fixed point $z = 0$ is stable (resp., unstable) for $\mu < 0$ (resp., $\mu > 0$). Such a bifurcation is commonly called a *Hopf bifurcation* or more accurately a *Poincaré–Andronov–Hopf bifurcation*. Henri Poincaré made the seminal contribution and imagined the notion of a limit cycle, and Andronov and Hopf developed the idea with improvements such as it gained a wider audience.

The linear system (4.1) is the germ of the normal form describing the bifurcation. As discussed in Chapter 3, it must be complemented by nonlinear terms to saturate the instability. The simplest expression for the normal form is

$$\dot{z} = (\mu + i\omega)z - \epsilon z|z|^2 = (\mu + i\omega - \epsilon|z|^2)z \qquad (4.2)$$

with $z \in \mathbb{C}$, $\epsilon = \pm 1$, and $\mu, \omega \in \mathbb{R}$. Since $\omega \neq 0$, there is still only one fixed point at $z = 0$. Note that Eq. (4.2) has rotational symmetry as it is invariant under $z \rightarrow ze^{i\theta}$. This is related to the oscillatory nature of the emerging solution.

As with the pitchfork bifurcation, the nonlinear term has two possible signs, and it is not possible to go from one case to the other through a change of variables. This leads to two different scenarios: the supercritical Hopf bifurcation ($\epsilon = 1$) and the subcritical Hopf bifurcation ($\epsilon = -1$), which requires higher-order terms to ensure the global stability of the system when $\mu > 0$.

Alternate formulations of Eq. (4.2) help us to gain insight into its solutions and their stability using the results of Chapter 2. Since physical systems are usually described with

https://doi.org/10.1515/9783110677874-004

real variables, we may want to describe the dynamics in the real plane using the coordinates x and y such that $z = x + iy$. Equation (4.2) can then be rewritten as

$$\begin{cases} \dot{x} = \mu x - \omega y - \epsilon x(x^2 + y^2), \\ \dot{y} = \mu y + \omega x - \epsilon y(x^2 + y^2). \end{cases}$$

The linearization in $(0,0)$ gives

$$\mathcal{L}|_{(0,0)} = \begin{bmatrix} \mu & -\omega \\ \omega & \mu \end{bmatrix},$$

from which we deduce $\Delta = \mu^2 + \omega^2 > 0$ and $T = 2\mu$ with the notations used in Section 2.1.5. Consequently, the fixed point $(0,0)$ is stable for $\mu < 0$ and unstable for $\mu > 0$. In this case, in fact, the eigenvalues of the Jacobian are easily computed to be $\lambda = (\mu \pm i\omega)$, which is not surprising given (4.1). At the bifurcation ($\mu = 0$) the behavior around the fixed point changes from a convergent spiral to a divergent spiral.

To better identify the nonzero solution emerging from the bifurcation $\mu > 0$, we express the complex variable z in polar coordinates ($z = re^{i\theta}$), rewriting (4.2) in terms of modulus and phase variables:

$$\begin{cases} \dot{r} = \mu r - er^3 = (\mu - er^2)r, \\ \dot{\theta} = \omega, \end{cases} \tag{4.3}$$

with the benefit of having two uncoupled one-dimensional time evolutions for r and θ. The phase θ increases uniformly with time, describing uniform rotation around the origin ($\theta = \omega t + \theta_0$), and does not carry essential information.

Let us first consider the simpler case $\epsilon = 1$. With the exception that r can only be positive or null, the differential equation for r is formally the same as for the one-dimensional pitchfork bifurcation, allowing us to build on the results in Section 3.3.3.

For $\mu < 0$, there is a single constant solution $r = 0$ corresponding to the stable fixed point at the origin. At $\mu = 0$, it destabilizes to a stable nonzero constant solution $r = r_0 = \sqrt{\mu}$, which exists only for positive μ (see Figure 4.1(a)). The motion occurs along a circle of radius r and is periodic with period $T = 2\pi/\omega$. This is our first example of a new type of invariant set, a *limit cycle*: starting from an initial condition located on the cycle, the system will remain on it forever.

This type of bifurcation is termed a *supercritical* Hopf bifurcation, because the emerging periodic solution exists beyond the bifurcation. Its bifurcation diagram is shown in Figure 4.2, in a three-dimensional perspective to account for the two-dimensional nature of the dynamics.

Let us now turn to the case $\epsilon = -1$. As in the pitchfork bifurcation, the normal form (4.3) diverges to infinity when $\mu > 0$ because $\dot{r} > 0$ for all r. We have then to continue the expansion until stabilizing higher-order terms are obtained. In fact, we

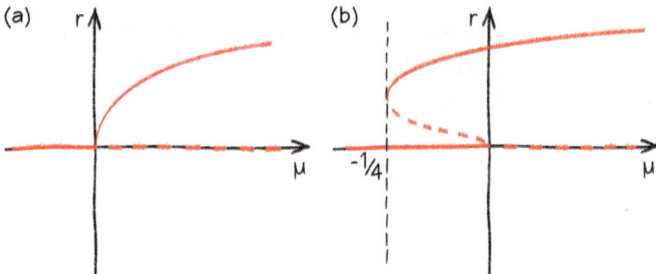

Figure 4.1: Bifurcation diagram of the modulus r of (a) supercritical and (b) subcritical Hopf bifurcations.

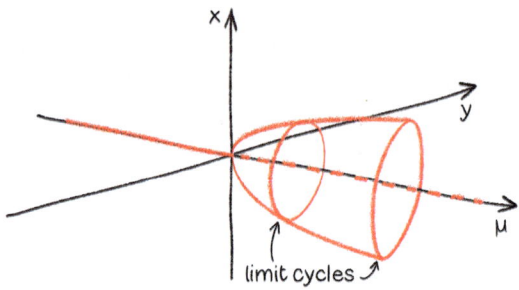

Figure 4.2: Bifurcation diagram of a supercritical Hopf bifurcation in a three-dimensional representation taking into account the two-dimensional nature of oscillations.

can show that in this case, it is always possible to carry out a change of variable so that the governing equation for the radial variable is

$$\dot{r} = \mu r + r^3 - r^5 = (\mu + r^2 - r^4)r. \tag{4.4}$$

where we recognize the normal form of the subcritical pitchfork bifurcation. For each of the strictly positive solutions of the bifurcation diagram of the latter (Section 3.3.3.b), there is a corresponding periodic solution in addition to the zero solution associated with the fixed point at the origin, as shown in Fig. 4.1(b).

Thus, Eq. (4.4), together with the phase equation $\dot{\theta} = \omega$, defines the subcritical Hopf bifurcation. An important difference of this system compared to the supercritical case is that the periodic solution emerging from the Hopf bifurcation is observed for $\mu < 0$. It thus coexists with the stable fixed point and encircles it (Fig. 4.1(b)). Because stable and unstable solutions alternate along the r-axis, the bifurcating periodic solution is unstable over its entire domain of existence.

The important new ingredient of the subcritical Hopf bifurcation is the existence of a stable periodic solution far away from the bifurcating fixed point. This stable solution is an attractor for the system and prevents it from escaping to infinity when the fixed point has become unstable. At $\mu = -1/4$, it is created together with the unstable periodic solution in a saddle-node bifurcation, not of fixed points as in the pitchfork bifurcation, but of closed orbits (Fig. 4.1(b)).

The stable periodic solution is present for any $\mu > -1/4$. For $-1/4 \leq \mu < 0$, it coexists with the stable fixed point (Fig. 4.1(b)), and hence we have bistability: the system can be either stationary or oscillating depending on whether the system starts from inside or outside the unstable periodic orbit, which is a separatrix between the two stable solutions. For $\mu > 0$, the stable periodic solution is the only attractor: all trajectories converge to it.

Note that when μ is continuously increased and the bifurcation point is crossed, the transition between the stable fixed point and the stable periodic solution is discontinuous: there is a sudden jump to the only remaining attractor, the stable limit cycle branch. This phenomenon is sometimes called a hard excitation, as large amplitude oscillations appear suddenly without any warning sign.

Another consequence of the bifurcation diagram of the subcritical Hopf bifurcation is a hysteresis phenomenon, as has been discussed for the pitchfork bifurcation (Section 3.3.3). After the system has jumped on the stable oscillating branch once the subcritical Hopf bifurcation at $\mu = 0$ has been crossed, it will remain on that branch if μ is decreased, until this branch disappears at the saddle-node bifurcation at $\mu = -1/4$, and then falls back on the fixed point solution. Thus the observation of bistability between stationary and oscillating solutions is often associated with a subcritical Hopf bifurcation.

4.1.2 Limit cycles

4.1.2.a Definition
The closed orbit solution

$$r = r_0 = \sqrt{\mu}, \quad \theta = \omega t + \theta_0 \tag{4.5}$$

of the supercritical Hopf normal form (4.2) is our first example of a *limit cycle*: a closed orbit, which is isolated in the phase space and which attracts neighboring trajectories either as $t \to +\infty$ (stable limit cycle) or as $t \to -\infty$ (unstable limit cycle). It is a closed orbit because the system goes back to any initial condition (r_0, θ_0) after exactly one period $T = 2\pi/\omega$. Away from the bifurcation, it is isolated in the sense that there is no other invariant set in its neighborhood, be it a closed orbit or the fixed point $r = 0$. Moreover, it is a stable limit cycle as $\lim_{t \to \infty} r(t) = r_0$.

Limit cycles only make sense for nonlinear dissipative systems. Linear systems are scale invariant (any rescaled solution is also a solution), and thus their closed orbits form a continuum. In nonlinear conservative systems, there is also a continuum of closed orbits surrounding a center, and there is again a continuum of them, nested within each other, as for the simple pendulum (Fig. 1.7 of Chapter 1). In both cases the amplitude of the oscillations is fixed by the initial condition. On the contrary, the trajectory and thus the amplitude and frequency of a limit cycle are intrinsic properties of the system and do not depend on the initial condition. For example, the amplitude and frequency of the limit cycle given by (4.5) are respectively $r = r_0$ and $\omega/2\pi$.

A limit cycle is a new example of an invariant set: once the system is on the closed orbit, it remains on it forever, and all points in the limit cycle belong to the orbits of each other. Consider the flow ϕ_τ such that $\phi_\tau(X(t_0)) = X(t_0 + \tau)$; then a limit cycle Λ satisfies $\phi_\tau(\Lambda) = \Lambda$ for all τ. Over a common time T, defined as the period of the limit cycle, all limit cycle points are mapped to themselves: $\phi_T(X) = X$ for all $X \in \Lambda$. Consequently, points on a limit cycle can be parameterized by a phase variable $\varphi \in S^1$.

4.1.2.b Poincaré–Bendixson theorem

As limit cycles can arise from the destabilization of a fixed point through a Hopf bifurcation, they are a generic dynamical behavior. They are even more important as it can be shown that fixed points and limit cycles are the only two possible asymptotic behaviors in a two-dimensional system whose dynamics is bounded in a finite region of the phase space. In this section, we discuss the Poincaré–Bendixson theorem, which makes this statement precise and reads as follows.

Consider an autonomous flow $\frac{dX}{dt} = F(X)$ with $X \in \mathbb{R}^2$ and continuously differentiable F. Assume that there is a closed bounded subset $\mathcal{R}$ of $\mathbb{R}^2$ that does not contain any fixed point. The theorem states that if there exists a trajectory confined in $\mathcal{R}$, then $\mathcal{R}$ contains a limit cycle, and the trajectory converges to it.

A simple way to ensure that there exists a trajectory confined in $\mathcal{R}$ is if $\mathcal{R}$ is a trapping region such that $\phi_t(\mathcal{R}) \subset \mathcal{R}$ for all $t > 0$. This property is guaranteed if on the boundary of $\mathcal{R}$, the vector field F is always directed toward the interior of $\mathcal{R}$ ($F \cdot n_{out} < 0$, where n_{out} is the outward normal to the boundary; see Fig. 4.3). Then all orbits starting in $\mathcal{R}$ remain in it and by the theorem wrap up around a limit cycle.

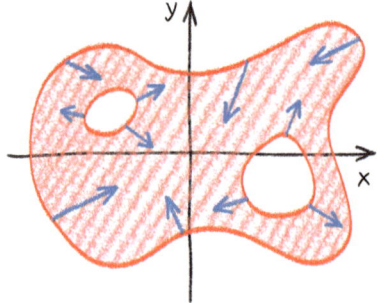

Figure 4.3: Illustration of the Poincaré–Bendixson theorem. If we can draw a set like the hatched one, containing no fixed points and such that on all the boundaries of this set the flow (represented by the blue arrows) goes *inside* the set, then the trajectories are trapped in this set, and there is a limit cycle in the hatched area.

A sketch of the proof is as follows. If there is an orbit confined in $\mathcal{R}$, then it will come back to previously visited points arbitrarily closely and arbitrarily many times, and hence there are recurrent points in $\mathcal{R}$ (Section 1.2.4). Let $\mathbf{X}_0 = \mathbf{X}(t_0)$ be one such recurrent point. Since $\mathcal{R}$ contains no fixed point, $\mathbf{F}(\mathbf{X}_0) \neq \mathbf{0}$. In a neighborhood of $\mathbf{X}_0$, this allows us to fix a curve $y(s)$ thats goes through $\mathbf{X}_0$ with $y(s_0) = \mathbf{X}_0$ and is everywhere transverse to the vector field $\mathbf{F}(\mathbf{X})$ (see Fig. 4.4). Since $\mathbf{X}_0$ is recurrent, the curve is crossed infinitely many times. Let us consider $\mathbf{X}_1 = \mathbf{X}(t_1) = y(s_1)$ as the first next intersection of the orbit $\{\phi_t(X_0)\}$ with $y(s)$, and assume that $\mathbf{X}_1 \neq \mathbf{X}_0$.

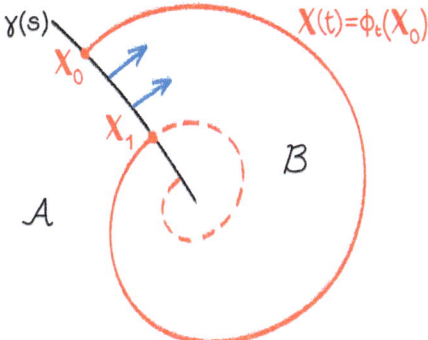

Figure 4.4: Proof of Poincaré–Bendixson theorem. $\mathbf{X}_0$ is a recurrent point, and $y(s)$ is a curve transverse to the flow at this point. After leaving $\mathbf{X}_0$, the orbit $\{\phi_t(\mathbf{X}_0)\}$ crosses again $y(s)$ in $\mathbf{X}_1$. The closed curve obtained by combining the orbit between $\mathbf{X}_0$ and $\mathbf{X}_1$ with the curve $y(s)$ between the same two points divides the plane into two disjoint regions $\mathcal{A}$ and $\mathcal{B}$.

Now we consider the (nonempty) closed curve formed by the arc of orbit $\{\phi_t(\mathbf{X}_0); 0 \leq t \leq t_1 - t_0\}$ and by the curve arc $y_{01} = \{y(s); s_0 \leq s \leq s_1\}$, which connect $\mathbf{X}_0$ to $\mathbf{X}_1$, and vice versa. By the Jordan curve theorem this closed curve divides the plane into interior and exterior regions. Since the $\{\phi_t(\mathbf{X}_0)\}$ orbit cannot cross itself (there is no fixed point in $\mathcal{R}$), the system can only transit from one region to the other by crossing the arc y_{01}. However, $y(s)$ is everywhere transverse to the vector field $\mathbf{F}(\mathbf{X})$ so that the flow can only drive the system from one of the two regions (call it $\mathcal{A}$) to the other region (call it $\mathcal{B}$), but not the converse. This implies that the orbit $\{\mathbf{X}(t); t > t_1\}$ remains forever in $\mathcal{B}$, which however precludes that this orbit returns arbitrarily close to $\mathbf{X}_0 = \mathbf{X}(t_0)$ since $\mathbf{X}_0$ can only be approached from $\mathcal{A}$. Hence $\mathbf{X}_0$ would not be a recurrent point, contradicting our initial assumption. The only possibility is to remove the assumption $\mathbf{X}_1 \neq \mathbf{X}_0$, implying that $\mathbf{X}_0$ and $\mathbf{X}_1$ are the same point along a closed orbit. In this case the recurrence is exact: there exists T such that $\phi_T(\mathbf{X}_0) = \mathbf{X}_0$, which thus belongs to a limit cycle.

We conclude that a recurrent point that is not a fixed point and is enclosed in a trapping region necessarily belongs to a limit cycle. Thus the only two possible asymptotic behaviors in two-dimensional systems are stationary solutions and periodic solutions (limit cycles). Note how the fact that two orbits cannot cross each other (nonintersection theorem) is essential to the Poincaré–Bendixson theorem. Actually, there may be several limit cycles in the trapping region, as is shown by the example of the subcritical Hopf bifurcation.

4.1.3 Example: The Brusselator

Let us apply those theoretical considerations on a practical example: a model of oscillating chemical reaction. Indeed, spectacular examples of oscillating dynamics are provided by oscillating chemical reactions such as the Belousov–Zhabotinsky reaction. Although we could expect that a chemical reaction always converges to equilibrium because of the decrease in free energy, it was clarified by I. Prigogine[1] and collaborators that oscillations are possible in systems that are maintained far from equilibrium by a constant supply of fresh reagents (Prigogine and Lefever, 1968).

Most of the time, realistic reactions involve many intermediate species and elementary reactions, which makes the formulation of the dynamical system difficult. Because of this complexity, a theoretical approach consists in the search of minimal models that lead to an oscillating dynamics. The *"Brusselator"* is a kinetical model proposed by I. Prigogine and R. Lefever in 1968 (Prigogine and Lefever, 1968). They considered the global reaction

$$A + B \rightarrow C + D$$

comprising four steps of elementary reactions and implying two free intermediate species X and Y:

$$A \rightarrow X,$$
$$B + X \rightarrow Y + C,$$
$$2X + Y \rightarrow 3X,$$
$$X \rightarrow D.$$

The control parameters of the system are the concentrations $[A]$ (resp. $[B]$) of species A (resp. B), assumed to be maintained at constant values by a continuous supply. All kinetic rates are supposed to be equal to 1. We omit the square brackets in the concentration notation (i. e. $[X] = X$) in the following.

The mass action laws lead us to the following system of differential equations:

$$\frac{dX}{dt} = A - BX + X^2Y - X,$$

---

1 Ilya Prigogine (1917–2003) was a Belgium physicist and chemist who received the Nobel Prize in Chemistry in 1977 for his contributions to the thermodynamics of irreversible processes and especially to the theory of dissipative structures.

$$\frac{dY}{dt} = BX - X^2 Y,$$

$$\frac{dC}{dt} = BX,$$

$$\frac{dD}{dt} = X.$$

Note that (X, Y) form an independent subsystem because their time derivatives only depend on themselves. Then the time course of C and D can be determined from that of X. Thus we focus here on the two-dimensional system

$$\frac{dX}{dt} = A - (B + 1)X + X^2 Y, \tag{4.6a}$$

$$\frac{dY}{dt} = BX - X^2 Y. \tag{4.6b}$$

We can now proceed to a linear stability analysis of this system, as we have discussed in Chapter 2. First, the fixed points of the system are found by solving the set of equations

$$A = [(B + 1) - XY]X,$$
$$BX = X^2 Y.$$

The only solution of this system is the fixed point $(A > 0)$

$$(X^*, Y^*) = \left(A, \frac{B}{A} \right). \tag{4.7}$$

The Jacobian of system (4.6) for arbitrary values of X and Y is

$$\mathcal{L} = \begin{pmatrix} -(B + 1) + 2XY & X^2 \\ B - 2XY & -X^2 \end{pmatrix}. \tag{4.8}$$

The stability of the fixed point is obtained by replacing X and Y in (4.8) by their fixed point values (Eq. (4.7)) and studying the eigenvalues of the matrix

$$\mathcal{L}|_{(A, \frac{B}{A})} = \begin{pmatrix} B - 1 & A^2 \\ -B & -A^2 \end{pmatrix}.$$

The determinant of the Jacobian is $\Delta = A^2 > 0$, and its trace is $T = B - 1 - A^2$. Because $\Delta > 0$, a bifurcation with a single zero eigenvalue is impossible. Using the results of Section 2.1.5 of Chapter 2, we conclude that when $B < 1 + A^2$, the fixed point is stable, whereas when $B > 1 + A^2$, it is unstable. At the bifurcation the fixed point shifts from a stable spiral point to an unstable spiral point.

Additionally, the existence of a limit cycle when $B > 1 + A^2$ can be demonstrated by invoking the Poincaré–Bendixson theorem. Let us first examine the geometry of null-clines in the (X, Y) plane:

$$Y = \frac{(B+1)X - A}{X^2}, \tag{4.9a}$$

$$Y = \frac{B}{X}. \tag{4.9b}$$

The Y-nullcline is a strictly decreasing curve, whereas the X-nullcline has a maximum at $X = 2A/(B+1)$ (see Fig. 4.5(a)). The two curves intersect at the fixed point (X^*, Y^*), whose position depends on the value of the parameters.

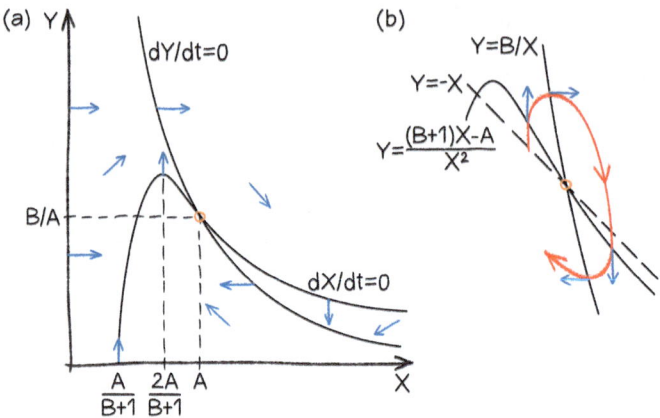

Figure 4.5: (a) Nullclines of Eq. (4.9) in the phase space (X, Y). As X and Y are concentrations of chemical species, only the quadrant $X > 0$ and $Y > 0$ is relevant. The general direction of the vector field is represented with blue arrows. (b) Enlargement in the vicinity of the fixed point in the unstable case, with representation of the nullclines, of the line of slope -1 and of a typical trajectory.

In Fig. 4.5 the general direction of the vector field in different regions of the phase space delimited by the nullclines has been represented. When the fixed point is unstable (i. e., when $B > 1 + A^2$), the X- and Y-nullclines are both steeper than a line of slope -1 as shown in Fig. 4.5(b). This forces the system to cross the Y-nullcline at a finite distance from the fixed point, preventing it from converging to the latter. There is no other alternative but to rotate around the fixed point forever.

Let us now apply the Poincaré–Bendixson theorem to the Brusselator system. For this, we need to find a trapping region such that the vector field on its boundary is always directed toward inside the region. First, we note that the nullclines divide the phase portrait into four areas, each corresponding to one general direction of the vector field, which can thus be qualified as pointing north-east (NE), NW, etc. To build the trapping region, we can pick a vertical line in the NE-pointing quadrant (where $\dot{X} > 0$), as well as in the SW-pointing and NW-pointing quadrants (where $\dot{X} < 0$) (see Fig. 4.6). To close the NW-pointing quadrant, the line $Y = 0$ does perfectly the job as we necessarily have $\dot{Y} > 0$. However, it is more delicate to close the boundary in the SE-pointing quadrant,

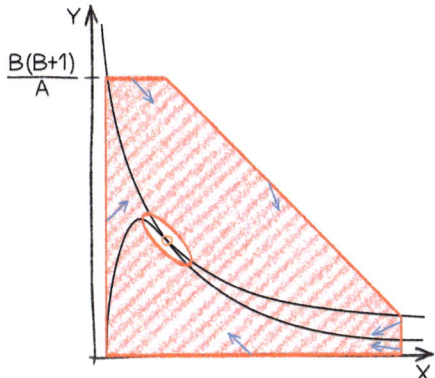

Figure 4.6: A trapping region for the Brusselator system. The outer boundary of this region is constructed in the following way, starting with the leftmost intersection of the X-nullcline with the X axis. (1) Go vertically upward until the Y-nullcline is crossed; (2) go horizontally to the right until $X = A$; (3) go south-east with a slope of $-1/2$ until the X-nullcline is crossed; (4) go vertically downward until the X axis crossed; (5) go back to starting point along the X axis. The inner boundary is obtained by drawing a small ellipse around the unstable fixed point.

as the boundary will have to go south-east, which is roughly the direction of the vector field.

Let us find a straight line with appropriate negative slope $\alpha = \frac{dY}{dX}$ and downward normal vector $\mathbf{N} = (\alpha, -1)$. The condition for the vector field to point inward the region is that the scalar product of $\mathbf{N}$ and $\mathbf{F} = (\dot{X}, \dot{Y})$, $\alpha\dot{X} - \dot{Y}$, is positive, and hence we must have

$$\alpha > \frac{\dot{Y}}{\dot{X}} \tag{4.10}$$

Now we have

$$\frac{\dot{Y}}{\dot{X}} = \frac{BX - X^2Y}{A - (B+1)X + X^2Y}$$
$$= -1 + \frac{A - X}{A - (B+1)X + X^2Y}. \tag{4.11}$$

The denominator of the fraction in (4.11) is always positive in the considered quadrant since it is equal to $\dot{X}$, and thus $\dot{Y}/\dot{X}$ is negative when $X > A$ (at the right of the fixed point), implying that $\frac{\dot{Y}}{\dot{X}} < -1$. It is thus sufficient that $\alpha > -1$. In Fig. 4.6, we used a straight line such that $\alpha = \frac{dY}{dX} = -\frac{1}{2}$ to close the outer boundary of the trapping region. The inner boundary is obtained by drawing a well-chosen small ellipse around the unstable fixed point.

In conclusion, we have found a trapping region containing no fixed point, and thus there exists a limit cycle inside it by the Poincaré–Bendixson theorem. Such a limit cy-

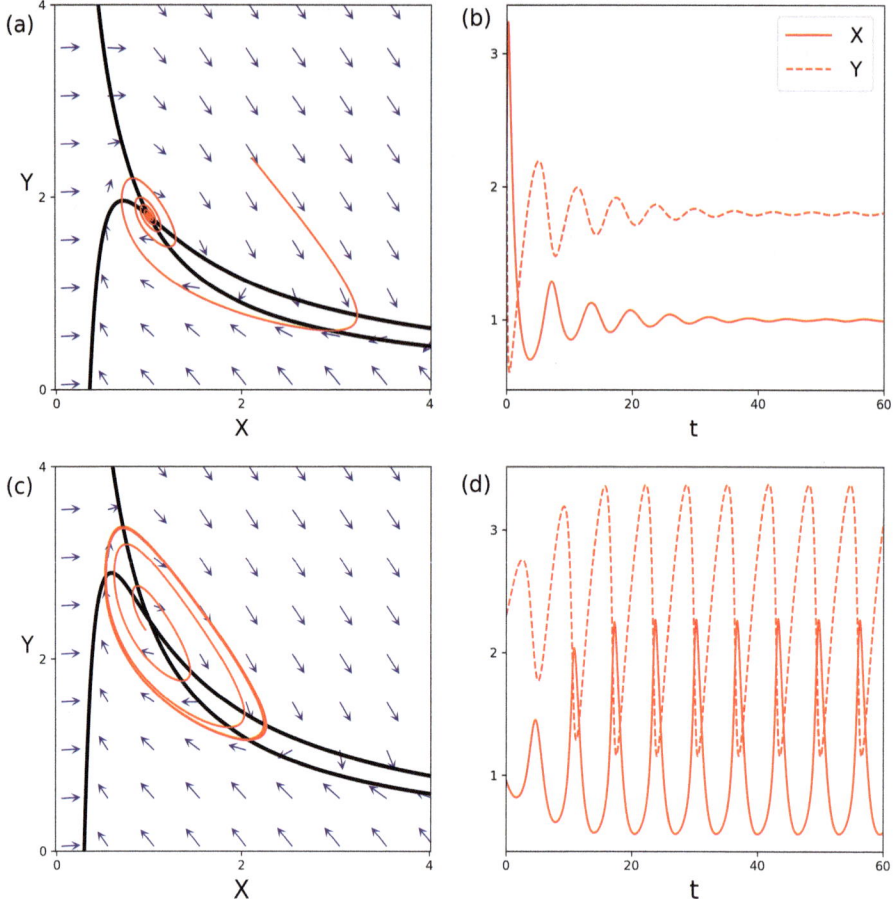

Figure 4.7: Numerical integration of Eqs. 4.6 for (a–b) $(A, B) = (1, 1.8)$ and (c–d) $(A, B) = (1, 2.4)$. (a) and (c): trajectories in the (X, Y) phase plane superimposed on the nullclines and the general directions of the vector field (the size of the vectors has been rescaled). (b) (resp., (d)) Temporal evolution of the concentrations of X (solid line) and Y (dashed line) corresponding to the trajectory plotted in (a) (resp., (c)).

cle is shown in Fig. 4.7c, which shows a trajectory obtained by numerical integration of Eqs. (4.6) in the (X, Y) phase plane. From an initial condition close to the fixed point, the trajectory spirals toward a closed orbit.

To illustrate the temporal behavior of the X and Y variables, we can consider the scenario where the concentration B is the control parameter, keeping the concentration A is constant. At low values of B, we observe a stationary state: the concentrations X and Y remain constant after a transient (Fig. 4.7b). When B increases, the corresponding fixed point destabilizes through a Hopf bifurcation leading to a limit cycle (Fig. 4.7c). The concentration in X and Y then display oscillations (Fig. 4.7d).

4.1.4 From weakly to strongly nonlinear oscillations: the van der Pol oscillator

A classical model giving rise to periodic oscillations is the van der Pol oscillator, named after Balthazar van der Pol, a Dutch engineer working at the Philips company in Eindhoven in the 1920s. At that time, there was a huge interest in building electrical devices generating oscillations at a prescribed frequency (Ginoux and Letellier, 2012).

Although the van der Pol equation was based on the physics of the vacuum tube used as a nonlinear element, it arises naturally as a simple second-order equation generating self-sustained oscillations. We start from a damped harmonic oscillator

$$\ddot{y} + \gamma\dot{y} + \omega^2 y = 0,$$

which is known to relax unconditionally to the fixed point solution $y(t) = 0$. Now let us assume that the damping γ is no longer a constant, and design it in such a way that energy is fed into the system around $y = 0$ (destabilizing the stationary solution) while it is dissipated far from the origin (keeping the system in a finite region of the phase space). The combination of these two processes leads the system to stabilize on a periodic solution.

A simple possible function $\gamma(y)$ satisfying these requirements is

$$\gamma(y) = y^2 - \mu$$

with $\mu > 0$. We obtain the nonlinear differential equation

$$\ddot{y} + (y^2 - \mu)\dot{y} + \omega^2 y = 0, \tag{4.12}$$

and we have:

- $\gamma(y) > 0$ when $|y| > \sqrt{\mu}$, i.e., for large amplitude oscillations, so that those oscillations are damped and their amplitude decreases;
- $\gamma(y) < 0$ when $|y| < \sqrt{\mu}$, i.e., for small amplitude oscillations, pushing the system away from the origin in the phase plane.

Let us rewrite Eq. (4.12) as a first-order system with the usual system of coordinates $Y_1 = y$ and $Y_2 = \dot{y}$:

$$\begin{cases} \dot{Y}_1 = Y_2, \\ \dot{Y}_2 = -\omega^2 Y_1 + (\mu - Y_1^2)Y_2. \end{cases} \tag{4.13}$$

The only fixed point is $\left(\begin{smallmatrix} 0 \\ 0 \end{smallmatrix}\right)$. The Jacobian matrix at this point is

$$\mathcal{L}|_{(0,0)} = \begin{pmatrix} 0 & -\omega^2 \\ 1 & \mu \end{pmatrix}.$$

Since $T = \mu$ and $\Delta = \omega^2 > 0$, the origin is a focus that destabilizes to a periodic solution through a Hopf bifurcation when $\mu > 0$.

4.1.5 Weakly nonlinear oscillations

When μ is small, we expect the system to remain in the neighborhood of the origin as discussed in Section 4.1.1. Accordingly, we first search for an approximate solution of the form $y_0(t) = a \cos \omega t$. Substituting this solution into (4.12), we obtain

$$-a\omega^2 \cos \omega t - a\omega(a^2 \cos^2 \omega t - \mu) \sin \omega t + a\omega^2 \cos \omega t = 0,$$

and thus

$$\left(\frac{a^2}{4} - \mu\right) \sin \omega t + \frac{a^2}{4} \sin 3\omega t = 0, \tag{4.14}$$

where we have used trigonometric relations to express the equation as a Fourier series whose coefficients should be zero. Equation (4.14) cannot be taken literally, because $y_0(t)$ is an ansatz that does not take into account the fact that the term $y^2\dot{y}$ in (4.12) generates terms oscillating with a pulsation $3\omega t$, which should be present in the solution. Consequently, we only keep the term of Eq. (4.14) oscillating at the fundamental frequency, consistent with the ansatz. This leads us to

$$y_0(t) = 2\mu^{1/2} \cos \omega t,$$

which, up to a term $\mathcal{O}(\mu^{3/2})$, satisfies (4.12). To carry the expansion a little further, we now try a solution of the form $y_1(t) = a \cos \omega t + b \cos(3\omega t + \phi)$ with $b = \mathcal{O}(\mu^{3/2})$. Exploiting again the smallness of the μ parameter, we find that at the leading order,

$$y(t) = 2\mu^{1/2} \cos \omega t - \frac{\mu^{3/2}}{4\omega} \sin 3\omega t. \tag{4.15}$$

Expression (4.15) illustrates in a simple way several universal properties of nonlinear oscillations:

– the amplitude of the oscillations is fixed by the control parameter μ and does not depend on initial conditions, which points to the existence of a limit cycle;
– the solution is no longer sinusoidal but has harmonic components (here at pulsation 3ω), which grow as the nonlinearity increases;
– the amplitude of the nonlinear oscillations grows as $\mu^{1/2}$ when moving away from the bifurcation, where μ is a typical control parameter that is zero at the bifurcation. Accordingly, the periodic solution grows very rapidly after the bifurcation.

There is a rigorous way to perform such a development, which implies introducing different time scales. A good introduction to such multiple scale analysis can be found in (Strogatz, 2018).

4.1.6 Strongly nonlinear oscillations

Let us now turn to the strongly nonlinear case ($\mu \gg 1$). In this case the amplitude of the solution of the van der Pol equation is not a relevant information, and thus we keep it finite by rescaling the equation. Simultaneously, we renormalize time to remove the unessential parameter ω.

Defining $t' = \omega t$, $x = y/\sqrt{\mu}$ leads to the following equation:

$$\ddot{x} + \tilde{\mu}(x^2 - 1)\dot{x} + x = 0, \tag{4.16}$$

where the derivatives are now relative to t' and $\tilde{\mu} = \mu/\omega$. Without ambiguity, we drop the tilde in the following. Equation (4.16) is in fact the classical formulation of the van der Pol equation, where μ only controls the strength of the nonlinearity rather than the distance to bifurcation. In this formulation, we see that the dissipation changes sign for $x = 1$, so that x will be $\mathcal{O}(1)$.

To understand the mechanisms generating the oscillations here, we will once again switch to a geometric description of the dynamics. Before that, we have to apply a useful transformation to Eq. (4.16). To this aim, note that it can be rewritten as

$$x = -\ddot{x} + \mu(1 - x^2)\dot{x} = -\frac{d}{dt}\left(\dot{x} + \mu\left(\frac{x^3}{3} - x\right)\right).$$

Denoting $S(x) = \frac{x^3}{3} - x$ and introducing the new variable $v = \dot{x}/\mu + S(x)$, which is $\mathcal{O}(1)$ as x, the van der Pol equation can be rewritten as the following first-order system:

$$\dot{x} = \mu(v - S(x)), \tag{4.17a}$$

$$\dot{v} = -\frac{x}{\mu}. \tag{4.17b}$$

The relevance of the curve $v = S(x)$ is now clear as it is one of the two nullclines of the system with x increasing (resp., decreasing) when $v > S(x)$ (resp., $v < S(x)$) (Fig. 4.8). The second nullcline is $x = 0$ with v increasing (resp., decreasing) when $x < 0$ (resp., $x > 0$).

Now the trick comes from the fact that Eqs. (4.17) are a *slow-fast system* where the two variables evolve on very different time scales in the $\mu \gg 1$ limit. In this case, the variable v can be considered as being most of the time frozen while x evolves rapidly to an equilibrium point given by $v = S(x)$, with the additional stability condition that $dS(x)/dx > 0$ (recall our discussion of one-dimensional systems in Section 1.3.1). Then there begins a period of slow evolution governed by Eq. (4.17b), where the system follows closely the curve $v = S(x)$ (Fig. 4.8).

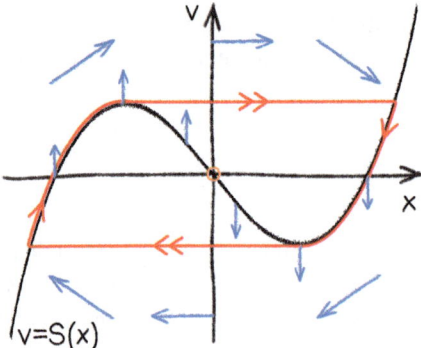

Figure 4.8: Phase portrait of the slow-fast system of Eqs. (4.17). $v = S(x)$ (resp. the y-axis) is the x-nullcline (resp., v-nullcline). The blue arrows indicate the general flow direction on the nullclines and between them. The limit cycle is plotted in red and alternates between a slow evolution (single arrow) along the $v = S(x)$ curve and a fast switch (double arrow) from one stable branch $S'(x) > 0$ to the other when one of the extrema of S is reached.

The oscillations are caused by the fact that given the geometry of the nullclines, the evolution of v will inexorably drive the system toward one of the turning points of the S-shaped curve $v = S(x)$ and then go beyond it, because $\dot{v}$ cannot change sign at these points. Since there is no stable equilibrium point in a neighborhood, the system is forced to undergo a phase of rapid evolution, jumping onto the only remaining branch of $S(x)$. In the process, it crosses the v-nullcline, and thus the dynamics of v changes direction. It does not take much time to realize that this scenario repeats itself forever.

Such oscillations are called *relaxation oscillations* because of the mechanism generating them. The basic ingredient is a bistable system with a control parameter that evolves on a slow time scale leading to the disappearance of the 1D fixed point that is followed, with a subsequent sudden relaxation to the other 1D fixed point.

4.2 Stability and bifurcation of periodic orbits

Since periodic orbits are important dynamical objects, we want to determine in which situation they will appear, disappear, or change their stability, as we did for fixed points in Chapters 2 and 3. This requires carrying out a linear stability analysis, which is much more complicated for a periodic orbit than for a fixed point, since we continuously move in the phase space. To circumvent this difficulty, we will use the fact that the motion is periodic, coming back to its initial condition after every period T_0. Hence any point $\mathbf{X}^*$ belonging to the orbit is a fixed point for the flow ϕ_{T_0} that evolves a state $\mathbf{X}(t)$ into $\mathbf{X}(t + T_0)$:

$$\phi_{T_0}(\mathbf{X}^*) = \mathbf{X}^*.$$

4.2.1 Stability of periodic orbits and their invariant manifolds

4.2.1.a Variational analysis around a periodic orbit

Consider a small perturbation $\delta\mathbf{X}_0$ of the fixed point $\mathbf{X}^*$. After a time T_0, it gets back not far from $\mathbf{X}^*$ at a position $\mathbf{X}^* + \delta\mathbf{X}_1 = \phi_{T_0}(\mathbf{X}^* + \delta\mathbf{X}_0)$. After kT_0, its position will be $\mathbf{X}^* + \delta\mathbf{X}_k = \phi_{T_0}(\mathbf{X}^* + \delta\mathbf{X}_{k-1})$. We can study the time evolution of the perturbation $\delta\mathbf{X}_k$ by linearizing this relation around the fixed point as in Section 2.2:

$$\mathbf{X}^* + \delta\mathbf{X}_{k+1} = \phi_{T_0}(\mathbf{X}^* + \delta\mathbf{X}_k) = \phi_{T_0}(\mathbf{X}^*) + \left(\frac{\partial\phi_{T_0}}{\partial\mathbf{X}}\right)\Big|_{\mathbf{X}^*}\delta\mathbf{X}_k = \mathbf{X}^* + \left(\frac{\partial\phi_{T_0}}{\partial\mathbf{X}}\right)\Big|_{\mathbf{X}^*}\delta\mathbf{X}_k,$$

and hence

$$\delta\mathbf{X}_{k+1} = \left(\frac{\partial\phi_{T_0}}{\partial\mathbf{X}}\right)\Big|_{\mathbf{X}^*}\delta\mathbf{X}_k = \mathcal{F}(\mathbf{X}^*)\delta\mathbf{X}_k. \tag{4.18}$$

We find again that the time evolution of the perturbation is governed by a linear equation, which is here a recurrence equation rather than a differential equation. Thus the stability of the orbit is determined by the eigenvalues of the matrix $\mathcal{F}$, called the *Floquet matrix*, and is the Jacobian of the flow ϕ_{T_0}. As described in Section 2.4, the Floquet matrix can be computed numerically by integrating the variational equation

$$\frac{d\delta\mathbf{X}(t)}{dt} = \left(\frac{\partial\mathbf{F}}{\partial\mathbf{X}}\right)\Big|_{\mathbf{X}=\mathbf{X}_{\mathrm{ref}}(t)}\delta\mathbf{X}(t)$$

along with the nonlinear equations $\dot{\mathbf{X}} = \mathbf{F}(\mathbf{X})$, so that the Floquet matrix $\mathcal{F}$ is the fundamental matrix solution $\mathbf{M}(T_0)$ of the linearized solution, which is also called the *monodromy matrix* in this context. The Floquet matrix typically depends on the base point $\mathbf{X}^*$ chosen, but its eigenvalues, which are a property of the periodic orbit, do not.

An important feature of the Floquet matrix is that it has always the velocity vector $\mathbf{F}(\mathbf{X})$ as an eigenvector with eigenvalue 1, expressing the invariance of the periodic orbit under time translation. Indeed, if a perturbation $\delta\mathbf{X}_f = \epsilon\mathbf{F}(\mathbf{X})$, then

$$\mathbf{X}(t) + \delta\mathbf{X}_f = \mathbf{X}(t) + \epsilon\mathbf{F}(\mathbf{X}) = \mathbf{X}(t) + \epsilon\dot{\mathbf{X}}(t) \simeq \mathbf{X}(t + \epsilon),$$

which indicates that the perturbation amounts to a time shift ϵ along the periodic orbit. Since the motion along the limit cycle has period T_0, we have $\mathbf{X}(t + \epsilon) = \mathbf{X}(t + \epsilon + T_0)$, and thus

$$\delta\mathbf{X}_f(t + T_0) \equiv \mathbf{M}(T_0)\delta\mathbf{X}_f(t) = \delta\mathbf{X}_f(t),$$

showing that $\delta\mathbf{X}_f$ is an eigenvector of $\mathbf{M}(T_0)$ with eigenvalue 1.

Following the same approach as in Chapter 2, we assume for simplicity that the Floquet matrix $\mathcal{F}$ is diagonalizable and that there is a change-of-basis matrix $\mathbf{P}$ such that in the new basis with coordinates $\delta\mathbf{Y} = \mathbf{P}\delta\mathbf{X}$, the Floquet matrix

$$\mathcal{F}_Y = \mathbf{P}\mathcal{F}\mathbf{P}^{-1}$$

is diagonal. Thus an initial perturbation δY_0 will grow after k periods into

$$
\delta Y_k = \begin{pmatrix} \lambda_1^k & 0 & \cdots & & 0 \\ 0 & \ddots & & & \\ \vdots & & \lambda_{n-1}^k & 0 \\ 0 & & 0 & 1 \end{pmatrix} \delta Y_0,
$$

where the λ_i are the eigenvalues of $\mathcal{F}$ in addition to the trivial eigenvalue 1 associated with the velocity vector $\mathbf{F}(\mathbf{X})$, also called the *Floquet multipliers*. We then conclude that

- If $|\lambda_i| < 1$ for all i, then the periodic orbit is stable because all $\lambda_i^k \to 0$ as $k \to \infty$. The periodic orbit is a limit cycle that attracts all neighboring trajectories and captures the asymptotic behavior.
- If there exists at least one λ_j such that $|\lambda_j| > 1$, then the periodic orbit is unstable because $|\lambda_j|^k \to \infty$ as $k \to \infty$. Note that a periodic orbit can be unstable in several directions simultaneously. The system will remain on the periodic orbit if the initial condition belongs to it ($\delta Y = 0$) but will diverge from it if there is the slightest perturbation along an unstable direction.
- If there is at least one eigenvalue λ_j such that $|\lambda_j| = 1$, then the periodic orbit is not structurally stable: it experiences a bifurcation. As we will see below, there are three different cases depending on whether the bifurcating eigenvalue is 1, −1, or a complex number e^{ia} (then the complex conjugate e^{-ia} is also an eigenvalue).

4.2.1.b Floquet multipliers and Poincaré map

Since our analysis indicates that stability is determined by only $(n-1)$ eigenvectors that are transverse with the vector field $\mathbf{F}(\mathbf{X})$, it is tempting to connect our results to the concept of a Poincaré section, which is an $(n-1)$-dimensional object transverse to the vector field (Section 1.4.1). Indeed, as we will demonstrate below, for any choice of surface for the Poincaré section (as long as the surface is transverse to the flow), the eigenvalues λ_i of the Floquet matrix are also the eigenvalues of the linearization around $\mathbf{X}^*$ of the Poincaré map $\mathcal{P}$, which maps a point in the Poincaré section to the first intersection of its orbit with the section, going in the chosen direction.

To see this, consider Σ, the tangent plane to the section surface at the base point $\mathbf{X}^*$, defined by the equation $\mathbf{N}.\delta\mathbf{X}$, where $\mathbf{N}$ is the normal vector to the section surface. For an arbitrary perturbation $\delta\mathbf{X}$, we define its projection on Σ parallel to $\mathbf{F}$ as $\mathbf{P}\delta\mathbf{X} = \delta\mathbf{X} - (\mathbf{N}.\delta\mathbf{X})/(\mathbf{N}.\mathbf{F})\mathbf{F}$, which satisfies the equation $\mathbf{N}.\mathbf{P}\delta\mathbf{X} = 0$ by construction (see Fig. 4.9). In other terms, the projection amounts to removing the component of a vector along $\mathbf{F}$. In particular, $\mathbf{PF} = 0$.

Because we work in a close neighborhood of the periodic point $\mathbf{X}^*$, projecting a perturbation $\delta\mathbf{X}$ on the Σ plane in parallel to $\mathbf{F}$ corresponds to considering the trajectory going through $\mathbf{X}^* + \delta\mathbf{X}$, whose tangent vector is parallel to $\mathbf{F}$ to first order, and following it backward or forward in time until it crosses Σ.

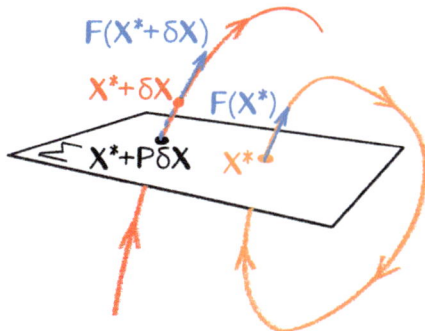

Figure 4.9: Illustration of the action of the projector **P**.

The tangent map $\mathcal{T} = (\partial\mathcal{P}/\partial\mathbf{X})$ of the Poincaré map is given by $\mathcal{T} = \mathbf{P}\mathcal{F}$. Indeed, starting from an initial perturbation $\delta\mathbf{X} \in \Sigma$, we can evolve it over one period by applying the Floquet matrix $\mathcal{F}$. Then we can find the intersection with Σ by following the trajectory forward or backward in time until it crosses Σ, which is exactly what the projector **P** does (see Fig. 4.9).

Consider now an eigenvector $\mathbf{V}_i$ of the Floquet matrix (see Fig. 4.10) such that $\mathcal{F}\mathbf{V}_i = \lambda_i\mathbf{V}_i$, and fix $\mathbf{W}_i = \mathbf{P}\mathbf{V}_i = \mathbf{V}_i - \beta\mathbf{F}$ so that $\mathbf{X}^* + \epsilon\mathbf{W}_i$ belongs to Σ and is on the same orbit as $\mathbf{X}^* + \epsilon\mathbf{V}_i$ (we do not need the precise value of β). The vector $\mathbf{W}_i$ is thus a combination of two eigenvectors of $\mathcal{F}$: $\mathbf{V}_i$ with eigenvalue λ_i and $\mathbf{F}$ with eigenvalue 1. Note also that $\mathbf{W}_i$ and $\mathbf{F}$ are two eigenvectors of the projector **P** with eigenvalues 1 and 0, respectively. Now

$$\mathcal{F}\mathbf{W}_i = \mathcal{F}\mathbf{V}_i - \beta\mathcal{F}\mathbf{F} = \lambda_i\mathbf{V}_i - \beta\mathbf{F} = \lambda_i\mathbf{W}_i + (\lambda_i - 1)\beta\mathbf{F},$$

and thus

$$\mathcal{T}\mathbf{W}_i = \mathbf{P}\mathcal{F}\mathbf{W}_i = \lambda_i\mathbf{W}_i, \tag{4.19}$$

showing that $\mathbf{W}_i$ is an eigenvector of $\mathcal{T}$, the tangent map at $\mathbf{X}^*$ of the Poincaré map $\mathcal{P}$, with eigenvalue λ_i (see Fig. 4.10). We now understand that the stability of an orbit only depends on its transverse stability, that is, on the response to perturbations transverse to the flow. Perturbations in the directions of the flow amount to time translation and stay on the periodic orbit.

4.2.1.c Invariant manifolds of a periodic orbit

As we did in Section 1.3.2 in the case of a fixed point, we can define the stable (W^s) and unstable (W^u) manifolds of a periodic orbit $\mathcal{O}$ as the sets of state space points whose orbit converges to the periodic orbit, respectively, as time $t \to \infty$ or $t \to -\infty$:

$$W^{s,u}(\mathcal{O}) = \left\{\mathbf{X} : \lim_{t\to\pm\infty} d(\phi_t(\mathbf{X}), \mathcal{O}) = 0\right\}, \tag{4.20}$$

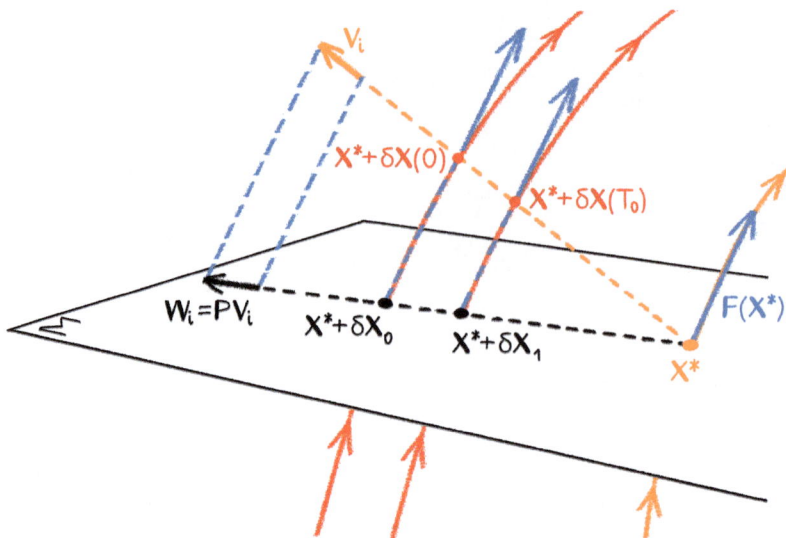

Figure 4.10: Behavior of a perturbation $\delta\mathbf{X}(t)$ along direction $\mathbf{V}_i$, an eigenvector of the Floquet matrix associated with an eigenvalue $|\lambda_i| < 1$ as well as the one of a perturbation $\delta\mathbf{X}_k$ in the Poincaré section surface. We can see that the perturbation in the section is reduced by the same amount as in the whole space. This is used to prove that the $n - 1$ eigenvalues of the Floquet matrix transverse to the flow direction are identical to the eigenvalues of the linearization of the Poincaré map in $\mathbf{X}^*$.

where $d(\mathbf{X}, \mathcal{O})$ represents the distance of a point $\mathbf{X}$ to the closest point on the orbit $\mathcal{O}$.

As discussed in Section 1.3.2.b, the stable and unstable manifolds are important objects, because an orbit contained in an invariant manifold remains in it by definition. Thus an invariant manifold forms a barrier that cannot be crossed transversely so that it structures the phase space.

To examine the geometry of invariant manifolds, it is again convenient to consider their intersection with a Poincaré section surface (Fig. 4.11). In the section surface the periodic orbit is represented by a fixed point $\mathbf{X}^*$ of the Poincaré map satisfying $\mathcal{P}(\mathbf{X}^*) = \mathbf{X}^*$. The intersections of the global invariant manifolds of the orbit with the Poincaré section surface coincide with the invariant manifolds of the fixed point $\mathbf{X}^*$ with respect to the Poincaré map $\mathcal{P}$. These consist of the points such that $\mathcal{P}(\mathbf{X}) \to \mathbf{X}^*$, either as $t \to \infty$ (stable manifold $W_{\mathcal{P}}^s$) or as $t \to -\infty$ (unstable manifold $W_{\mathcal{P}}^u$):

$$W_{\mathcal{P}}^{s,u}(\mathbf{X}^*) = \left\{ \mathbf{X} \in \Sigma : \lim_{k \to \pm\infty} \mathcal{P}^k(\mathbf{X}) = \mathbf{X}^* \right\}. \tag{4.21}$$

In a neighborhood of $\mathbf{X}^*$ the Poincaré section surface can be approximated by its tangent plane: we can approximate the first return map $\mathcal{P}$ by its tangent map $\mathcal{T}$ at the fixed point. Since $\mathcal{T}$ is a linear map, it can be diagonalized, and the linear span of eigendirections corresponding to eigenvalues of negative real part (resp., positive real part) provide a tangent space to the intersection of the stable (resp., unstable) manifold with the

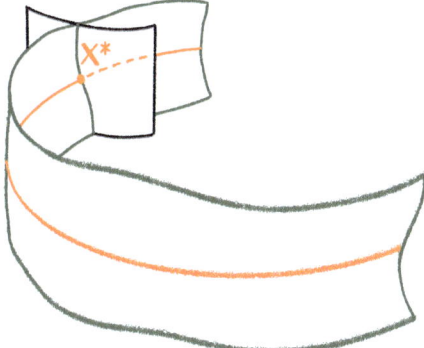

Figure 4.11: Intersection of an invariant manifold W (green surface) of a limit cycle (orange line) with a Poincaré surface (black surface). The green line in the Poincaré section surface is an invariant manifold $W_{\mathcal{P}}$ of the fixed point $\mathbf{X}^*$ with respect to the Poincaré map $\mathcal{P}$.

section surface in an infinitesimal neighborhood of $\mathbf{X}^*$. In fact, infinitesimal segments aligned around $\mathbf{X}^*$ along the eigendirections of the tangent map can be considered as germs of invariant curves in the Poincaré section, since the latter can be obtained from the former by indefinitely iterating the forward or backward map.

Interestingly, the dynamics along the stable manifold in the section surface is continuously squeezed toward the fixed point so that it can quickly be ignored. After a transient has died out, only the dynamics along the unstable manifold or along a possible *central* manifold (associated with an eigenvalue of modulus 1) remains. This can lead to a dramatic reduction of the complexity when those directions have low dimensionality.

The case where one or two eigenvalues have modulus one is particularly important because it is associated with perturbations that persist indefinitely, indicating in fact a degeneracy of the invariant solutions. This is the signature of a bifurcation of the periodic orbit, as discussed in Section 4.2.2.

4.2.2 Bifurcations of periodic orbits

In Section 3.3, we found that a fixed point of a one-dimensional flow is structurally stable when the eigenvalues of the Jacobian of the linearized flow at the fixed point have negative real parts.

Here we consider the conditions under which the fixed point of a map of a surface into itself (representing a periodic point) is structurally stable. Again, we take into account explicitly the dependence of the Poincaré map on a parameter μ and write the fixed point equation a:

$$\mathcal{P}(\mathbf{X}^*, \mu) = \mathbf{X}^*. \tag{4.22}$$

Assume that the control parameter is changed to $\mu + \delta\mu$. Then the fixed point is changed to $\mathbf{X}^* + \delta\mathbf{X}^*$, so that

$$\mathcal{P}(\mathbf{X}^* + \delta\mathbf{X}^*, \mu + \delta\mu) = \mathbf{X}^* + \delta\mathbf{X}^*. \tag{4.23}$$

Developing the left-hand side of (4.23) to first order, we find that the fixed point displacement $\delta\mathbf{X}^*$ satisfies

$$\left(\frac{\partial\mathcal{P}}{\partial\mathbf{X}} - \mathbb{1}\right)\delta\mathbf{X}^* = -\frac{\partial\mathcal{P}}{\partial\mu}\delta\mu. \tag{4.24}$$

This linear equation is solvable, and thus the periodic orbit is structurally stable as long as the Poincaré tangent map $\frac{\partial\mathcal{P}}{\partial\mathbf{X}}$ does not have 1 as an eigenvalue. For a map, the above condition is equivalent to imposing the absence of tangency around the fixed point of a flow, as discussed in Section 3.2.3. It is illustrated for a one-dimensional map in Fig. 4.12.

However, a bifurcation occurs not only when an invariant solution (here a periodic orbit) appears or disappears, but also when it changes its stability. A change of stability occurs whenever the modulus of an eigenvalue of the tangent map crosses 1, which marks the boundary between vanishing perturbations and indefinitely growing perturbations (see Section 4.2.1).

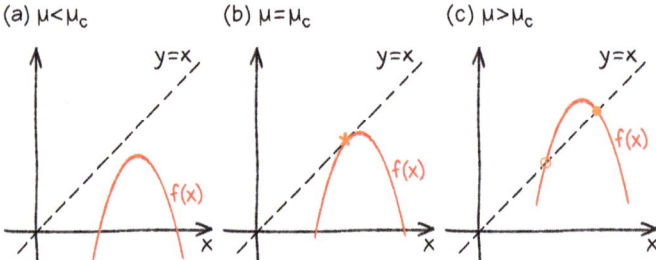

Figure 4.12: Bifurcation condition of Eq. (4.24) in the case of a one-dimensional map $f(x,\mu)$. When the parameter μ is tuned, the condition $f'(x) = 1$ leads to the emergence of a pair of periodic orbit in the Poincaré section.

For simplicity, here we will consider bifurcations experienced by a totally stable periodic orbit (i. e., all $|\lambda_i| < 1$ before the bifurcation). Then we have to distinguish two cases:

– A single real eigenvalue has modulus 1 at the bifurcation, and thus we have $\lambda_1 = \pm1$. We will see below that this corresponds to the saddle-node and the period-doubling bifurcations. In this case, we can approximate the Poincaré first return map with a one-dimensional map $x_{k+1} = f(x_k)$. We will take advantage of this below.

– A couple of complex conjugate eigenvalues $\lambda_{1,2} = e^{\pm i\theta}$ have modulus 1 at the bifurcation. In this case, we have to consider a two-dimensional return map since the unstable manifold is of dimension 2. The first return map then experiences a *Neimark–Sacker bifurcation*, with an invariant curve emerging in the Poincaré section around the fixed point associated with the periodic orbit. This bifurcation can be pictured as a Hopf bifurcation around a Hopf bifurcation.

4.2.2.a Saddle-node bifurcations of periodic orbits

Here we consider the case of a single real eigenvalue $\lambda_1 = +1$. In this case the relevant dynamics in the Poincaré map takes place in the eigenspace associated with λ_1 and can be modeled by a 1D iteration map $x_{k+1} = f(x_k)$ with $\frac{\partial f}{\partial x}(x^*, 0) = \lambda_1 = +1$. Without loss of generality, we assume that the bifurcation occurs at $\mu = 0$ and $x^* = 0$, so that

$$f(0,0) = 0, \quad f_x(0,0) = 1,$$

where we have used the notation $f_x = \partial f/\partial x$ as in Section 3.2.4. Reproducing the analysis followed in Chapter 3, we then consider a Taylor expansion of f with infinitesimally small x and μ, keeping only the lowest-order terms leading to a nontrivial equation. In the generic case where $f_\mu \neq 0$, with the adequate normalization of μ, we thus obtain

$$f(x, \mu) = \mu + x + \alpha x^2. \tag{4.25}$$

The fixed point equation $f(x^*, \mu) = x^*$ can then be rewritten as

$$\mu + \alpha x^2 = 0, \tag{4.26}$$

which has real solutions when $\mu/\alpha < 0$:

$$x_{\pm}^* = \pm \sqrt{\frac{-\mu}{\alpha}}. \tag{4.27}$$

We thus have a pair of periodic orbits on one side of the bifurcation (when $\mu/\alpha < 0$) and none on the other side, as illustrated by Fig. 4.12. Moreover, we see that since $\frac{\partial f}{\partial x} = 1 + 2\alpha x^* = 1 \pm 2\sqrt{-\alpha\mu}$, the two periodic solutions have opposite stabilities as their leading multipliers are larger and smaller than 1. This is obvious in Fig. 4.12, where we see that the slope of the graph of f is necessarily greater than 1 for one fixed point and smaller than 1 for the other.

This is the typical behavior of a saddle-node bifurcation, as we have encountered in Section 3.3.1, occurring here for a pair of periodic orbits, as illustrated in Fig. 4.13.

4.2.2.b Period-doubling bifurcations of periodic orbits

Here we consider the case where one Floquet multiplier is −1. Still considering that the bifurcation occurs for $\mu = 0$ and $x = 0$, we thus have

$$f(0,0) = 0, \quad f_x(0,0) = -1.$$

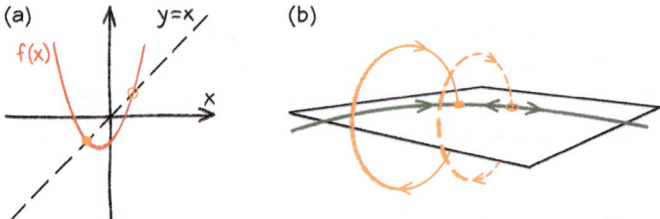

Figure 4.13: Saddle-node bifurcation of periodic orbits.(a) Fixed points of Eq. (4.25) when $a > 0$ and $\mu < 0$. (b) Corresponding periodic orbits.

Here the bifurcating orbit exists throughout the bifurcation. As we discussed for the transcritical and pitchfork bifurcations (Sections 3.2.4.b and 3.2.4.c), we can therefore choose the coordinates so that $f_\mu(0,0) = f_{\mu\mu}(0,0) = 0$.

In this case, the leading order in μ in the Taylor expansion of f is $f_{x\mu}$. We will thus consider the following expansion for small x and μ:

$$f(x, \mu) = -(1 + \mu)x + ax^2 = \lambda x + ax^2 \qquad (4.28)$$

with the eigenvalue $\lambda = -(1 + \mu)$ of this one-dimensional map crossing -1 when μ crosses 0.

We note that if the first return map has a Floquet multiplier of -1, then the doubly iterated first return map has a Floquet multiplier of $(-1)^2 = 1$ at the bifurcation and thus displays a saddle-node bifurcation with a pair of fixed points appearing in the bifurcation. Since these fixed points of the doubly iterated first return map are exchanged between them under the return map, they form a period-two orbit for the latter.

Let us verify this directly. If $f(x) = x(\lambda + ax)$, then the period-2 orbits of f satisfy

$$f^2(x) - x = x(\lambda + ax)(\lambda + ax(\lambda + ax)) - x = 0. \qquad (4.29)$$

Substituting $\lambda = -1 - \mu$ into Eq. (4.29), the leading order term is

$$f^2(x) - x \sim 2x(\mu - a^2x^2).$$

We see that in addition to the obvious solution $x = 0$, we have a pair of solutions

$$x_\pm^* = \pm\frac{\sqrt{\mu}}{|a|} \sim \mu^{\frac{1}{2}}, \qquad (4.30)$$

which only exist when $\mu > 0$, that is, when the period-1 orbit has become unstable ($\lambda = -1 - \mu$). These are fixed points for the doubly iterated map but make together a period-2 orbit for the first return map.

Thus the period-doubling bifurcation for f is a pitchfork bifurcation for f^2. In Section 3.2.4.c, we saw that a pitchfork bifurcation is associated with a symmetry breaking.

Here it is a symmetry in time that is broken: the iterates of f repeat only every other period instead of every period.

The largest Floquet multiplier of the period-2 orbit is $v = f_x(x_+^*)f_x(x_-^*) = 1-2\mu$ to leading order. Hence we find that the period-doubled orbit is stable beyond the bifurcation. The scenario of the period-doubling bifurcation is illustrated in Fig. 4.14.

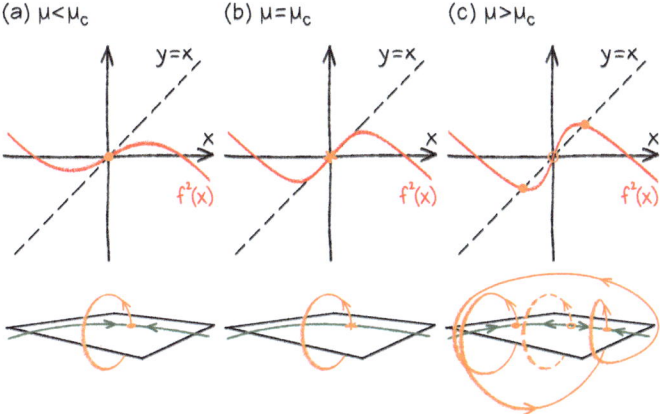

(a) $\mu<\mu_c$ (b) $\mu=\mu_c$ (c) $\mu>\mu_c$

Figure 4.14: Period-doubling bifurcation of a periodic orbit. Top row: second-return map f^2 (a) before, (b) at, and (c) after the bifurcation. The periodic orbit corresponding to $x^* = 0$ becomes unstable at the bifurcation simultaneously with the emergence of two stable fixed points. Those new fixed points of f^2 correspond to a period-2 orbit for the first-return map f. The corresponding limit cycles in the phase space are plotted in the bottom row. The fixed points $x_\pm^*$ of f^2 belong to a double-loop limit cycle, given that they are exchanged under the first return map.

4.2.2.c Neimark–Sacker bifurcation around a periodic orbit

When there is a couple of complex conjugate eigenvalues $\lambda_{1,2} = e^{\pm i\theta}$ whose modulus crosses 1 at the bifurcation, their eigendirections form a plane, and then we have to consider a two-dimensional first return map. This bifurcation, which is called a *Neimark–Sacker bifurcation*, shares several common points with the Hopf bifurcation we studied at the beginning of this chapter. Indeed, the 2D Poincaré map is then roughly equivalent to a rotation. Before the bifurcation, the nearby trajectories wrap around the orbit while converging toward it (Fig. 4.15a). After the bifurcation, the periodic orbit is unstable, and then the trajectories are attracted to a new attractor, which is an invariant closed curve in the Poincaré section. This curve can be considered as the intersection with the section plane of a new type of attractor, a torus (Fig. 4.15b), which we will study in Chapter 5.

The dynamics on the closed invariant curve can be quite complicated. Typically, it is periodic if θ is close to a rational number, and the dynamics in the Poincaré section consists in an alternation among a finite number of points. If θ is irrational, then the dy-

(a)   (b)

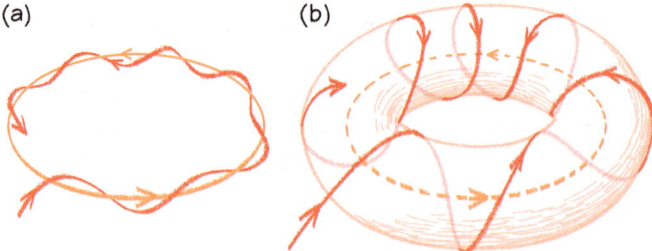

Figure 4.15: Illustration of the Neimark–Sacker bifurcation showing the behavior (a) before the bifurcation, with dampened oscillations around the periodic orbit, and (b) after the bifurcation, with the birth of an invariant curve and an invariant torus.

namics is quasi-periodic, and the system explores densely the torus without ever coming back to a previous point.

In both cases the invariant curve is homeomorphic to a circle, and thus the dynamics on it can be described by a map from a circle into itself. We will discuss the dynamics of such maps in Section 4.3.2 when we will study the synchronization of oscillators.

4.3 Driven oscillations and synchronization

In this chapter, we have so far considered autonomous self-sustained oscillations of systems that were isolated from the rest of the world. However, it is often the case that systems are subjected to some external driving. In this section, we study two important cases depending on whether the driven system is a self-sustained oscillator or not. First, we will study the periodic driving of damped nonlinear oscillators that do not display spontaneous oscillations, showing how the resonances of a damped oscillator are modified in the presence of nonlinearities. We will also consider the parametric instability, where the periodic modulation of a control parameter destabilizes a system. In a second part, we will deal with the important and complex case of a self-sustained oscillator driven by an external cycle, studying how oscillations can synchronize.

4.3.1 Driven oscillations

In this first part, we study briefly oscillations of a damped nonlinear oscillator subjected to a periodic driving. Two types of forcing are to be distinguished according to whether the external modulation directly influences the time evolution of the state variables (usually, this amounts to adding a driving term to the other terms of the differential equation) or a parameter of the equation. The latter case is called *parametric driving*.

The difference between those cases can be illustrated by two ways of making a swing (the children's game) oscillate. You can sit on it and periodically bend and stretch your legs, creating a torque that influences the swing oscillation. The system then responds at the same frequency as the movement of your legs. The motion of the legs can be modeled by adding a periodic term in the equation of motion, and the oscillations of the swing behave as for a usual forced oscillator. However, you can also stand on the swing and squat up and down periodically, a movement that can be modeled as a modulation of the distance of your center of gravity to the rotation axis. The swing then oscillates at half the driving frequency, which is the signature of a nonlinear response. This is related to the fact that parametric driving is intrinsically nonlinear, whereas the usual forcing can be considered linear.

4.3.1.a Nonlinear resonances in the Duffing oscillator

In the classical study of the resonance phenomenon in a driven harmonic oscillator, it is found that the amplitude of the response is maximum when the natural frequency of the oscillator and the frequency of the forcing coincide. Also, the response is less intense but also less selective as damping increases.

Here we will see how nonlinearity modifies this scenario. In particular, we will observe that the response can become bistable, with high and low amplitude responses coexisting for the same excitation frequency. To this aim, we consider a particular oscillator, the driven *Duffing oscillator*, governed by the equation

$$\ddot{x} + \gamma\dot{x} + \omega_0^2 x + \alpha x^3 = F\cos\omega t. \tag{4.31}$$

Following (Holmes and Rand, 1976), we consider the case where nonlinearity, damping, and forcing are all small. Denoting $\gamma = \varepsilon\gamma_1$, $\alpha = \varepsilon\alpha_1$, and $F = \varepsilon F_1$ with $\varepsilon \ll 1$, we rearrange Eq. (4.31) as follows:

$$\ddot{x} + \omega_0 x^2 = \epsilon(-\gamma_1\dot{x} - \alpha_1 x^3 + F_1\cos\omega t), \tag{4.32}$$

where we are close to resonance, with $\omega^2 - \omega_0^2 = \varepsilon\,\omega\delta\Omega$, i. e., $\varepsilon\,\delta\Omega \simeq \frac{\omega-\omega_0}{2}$. This approximation allows us to search solutions in the form of quasi-sinusoidal solutions $x(t) = u(t)\cos\omega t + v(t)\sin\omega t$, where u and v are slowly varying amplitudes that should converge to an equilibrium since we expect the forced solutions to have the same frequency as the driving term. An interesting approach is then to consider that u and v are the coordinates in a rotating frame:

$$\begin{pmatrix} x \\ \frac{\dot{x}}{\omega} \end{pmatrix} = \begin{pmatrix} \cos\omega t & \sin\omega t \\ -\sin\omega t & \cos\omega t \end{pmatrix} \begin{pmatrix} u \\ v \end{pmatrix}. \tag{4.33}$$

Inverting this relation, we obtain

$$u = x\cos\omega t - \frac{\dot{x}}{\omega}\sin\omega t, \tag{4.34a}$$

$$v = x\sin\omega t + \frac{\dot{x}}{\omega}\cos\omega t. \tag{4.34b}$$

By differentiating these two relations we obtain

$$\dot{u} = -\frac{\sin \omega t}{\omega}(\omega^2 x + \ddot{x}) = -\frac{\varepsilon \sin \omega t}{\omega}(\omega \delta \Omega x - \gamma_1 \dot{x} - \alpha_1 x^3 + F_1 \cos \omega t), \quad (4.35a)$$

$$\dot{v} = \frac{\cos \omega t}{\omega}(\omega^2 x + \ddot{x}) = \frac{\varepsilon \cos \omega t}{\omega}(\omega \delta \Omega x - \gamma_1 \dot{x} - \alpha_1 x^3 + F_1 \cos \omega t), \quad (4.35b)$$

where we see that u and v are constant when $\varepsilon = 0$ and that they are otherwise slowly varying. Substituting expressions (4.33) into (4.35), we take the advantage of the slow dynamics of u an v in Eqs. (4.35) and average the small driving terms in Eqs. (4.35) over one driving period, integrating them over $[0, \frac{2\pi}{\omega}]$. We thus obtain the following equations governing the slow evolution of u and v:

$$\dot{u} = \frac{\varepsilon}{2}(-\gamma_1 u - \delta \Omega' v), \quad (4.36a)$$

$$\dot{v} = \frac{\varepsilon}{2}\left(\delta \Omega' u - \gamma_1 v + \frac{F_1}{\omega}\right), \quad (4.36b)$$

where $\delta \Omega' = \delta \Omega - \frac{3\alpha_1}{4\omega}(u^2 + v^2)$. Interestingly, Eqs. (4.36) are linear except for the non-linear expression of $\delta \Omega'$, indicating a dependance of the frequency on amplitude. This frequency pulling is directly related to the nonlinearity in the Duffing equation.

Since here we are mostly interested in determining the amplitude of quasi-sinusoidal oscillations, we switch to the polar coordinates $r = \sqrt{u^2 + v^2}$ and $\varphi = \arctan(\frac{v}{u})$, whose time derivatives satisfy $r\dot{r} = u\dot{u} + v\dot{v}$ and $r^2\dot{\varphi} = u\dot{v} - v\dot{u}$. After the change of variable, we obtain the following equations:

$$\dot{r} = \frac{\varepsilon}{2}\left(-\gamma_1 r + \frac{F_1}{\omega}\sin \varphi\right), \quad (4.37a)$$

$$r\dot{\varphi} = \frac{\varepsilon}{2}\left(\left(\delta \Omega - \frac{3}{4}\frac{\alpha_1}{\omega}r^2\right)r + \frac{F_1}{\omega}\cos \varphi\right), \quad (4.37b)$$

and we search for fixed points of Eqs. (4.37), corresponding to a periodic solution at frequency ω with fixed amplitude and phase. Eliminating the phase φ between the two equations (4.37), we find that r^2 satisfies the following cubic equation with unknown X:

$$\frac{9}{16}\left(\frac{\alpha_1}{\omega}\right)^2 X^3 - \frac{3}{2}\frac{\alpha_1}{\omega}\delta \Omega X^2 + (\gamma_1^2 + \delta \Omega^2)X - \left(\frac{F_1}{\omega}\right)^2 = 0. \quad (4.38)$$

Multiplying Eq. (4.38) by $(\varepsilon \omega)^2$, we obtain a cubic equation dependent on the original parameters:

$$\frac{9}{16}\alpha^2 X^3 - \frac{3}{2}\alpha(\omega^2 - \omega_0^2)X^2 + (\gamma^2 \omega^2 + (\omega^2 - \omega_0^2)^2)X - F^2 = 0. \quad (4.39)$$

Depending on the value of the control parameters, there exist one or three solutions, indicating a bistable response: for the same forcing, we can observe a strong or weak oscillation. In the linear case ($\alpha = 0$), Eq. (4.39) is of first order, and there is only one solution. In the nonlinear case, the Cardano formulas for solving cubic equations can be used. Figure 4.16 shows examples of the resonance curves $r = r(\omega)$ for different values of the nonlinearity α. Alternatively, Eq. (4.38) can be viewed as a quadratic equation in $\delta\Omega$ with parameter $X = r^2$. It can be shown that the high- and low-amplitude oscillations are stable, while the intermediate solution that lies between them is unstable.

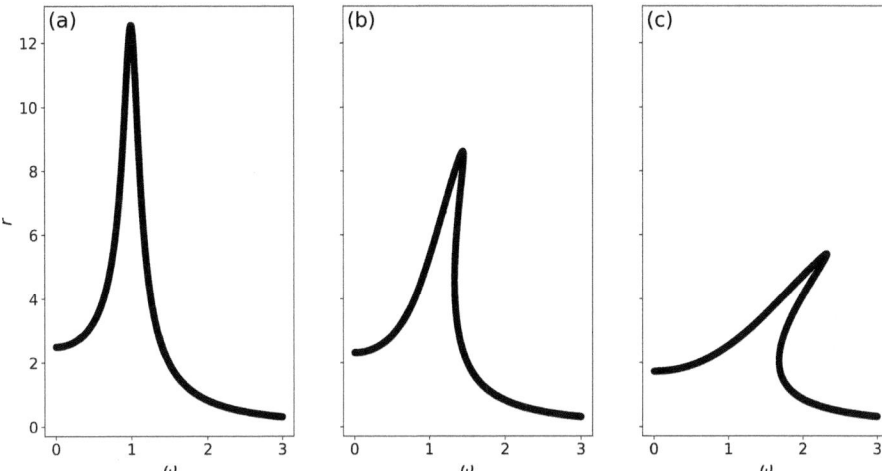

Figure 4.16: Dependence of the amplitude r of the forced oscillations on the driving frequency ω for different values of α, according to Eq. (4.39) for $\gamma = 0.2$, $F = 2.5$, and $\omega_0 = 1$ for (a) $\alpha = 0$, (b) $\alpha = 0.02$, and (c) $\alpha = 0.2$. Note that as $X = r^2 \geq 0$, only positive roots of Eq. (4.39) are used to plot the graph.

Note that the approach followed here may be applied to any weakly nonlinear oscillator of the form $\ddot{x} + \omega_0^2 x = \epsilon f(\dot{x}, x, t)$, where all nonlinear terms are contained in the small right-hand side term. In particular, we could have applied it to the classical formulation of the van der Pol oscillator given by Eq. (4.16) when μ is a small parameter. In this case, we would have found that the amplitude r obeys the equation

$$\dot{r} = \frac{\mu r}{2}\left(1 - \frac{r^2}{4}\right),$$

implying that $r = 2$ asymptotically. This is consistent with the first term of the solution (4.15) once the original scaling is restored.

4.3.1.b Parametric oscillations

We now study the case of parametric driving, where a control parameter of a differential equation is modulated. To illustrate this phenomenon, we consider a weakly damped pendulum whose resonance frequency is subjected to a small periodic modulation at a frequency close to the double the resonance frequency ($\omega \sim \omega_0$):

$$\ddot{x} + \gamma\dot{x} + \omega_0^2(1 + \mu\cos 2\omega t)x = 0. \tag{4.40}$$

Besides the swing we mentioned in the introduction of Section 4.3, a famous example of such a system is the swinging incense burner of the cathedral of Santiago de Compostela, whose length is periodically modulated by pulling and loosening the rope that supports it, and which displays impressive oscillations.

We recognize that Eq. (4.40) can be put in a form similar to (4.32), and thus we can follow the same approach, assuming again that $\omega^2 - \omega_0^2 = \varepsilon\omega\delta\Omega$, $\mu = \varepsilon\mu_1$, and $\gamma = \varepsilon\gamma_1$ and that $x(t) = u(t)\cos\omega t + v(t)\sin\omega t$. Here we obtain the following equations for the two amplitude components u and v:

$$\dot{u} = \frac{\varepsilon}{2}\left(-\gamma_1 u - \left(\delta\Omega + \frac{\mu_1\omega_0^2}{2\omega}\right)v\right), \tag{4.41a}$$

$$\dot{v} = \frac{\varepsilon}{2}\left(\left(\delta\Omega - \frac{\mu_1\omega_0^2}{2\omega}\right)u - \gamma_1 v\right). \tag{4.41b}$$

These are homogeneous linear equations admitting the origin as the only fixed point, and hence oscillations will appear when this fixed point becomes unstable. Computing the characteristic polynomial of the matrix associated with Eqs. (4.41), we find that it has a positive real eigenvalue when

$$\mu_1^2 \geq \frac{4\omega^2}{\omega_0^4}(\gamma_1^2 + \delta\Omega^2).$$

To get back to the parameters of the original equation (4.40), we denote $\delta\omega = \omega - \omega_0$ and assume that $\gamma, \delta\omega \ll \omega_0$, so that the equation translates into

$$\mu \geq \frac{2}{\omega_0}\sqrt{\gamma^2 + 4\delta\omega^2}, \tag{4.42}$$

which is the condition for the existence of parametric oscillations. Our linear analysis in the vicinity of the fixed point breaks down when the fixed point becomes unstable, but here we are only interested in finding the regions where the oscillations develop.

Figure 4.17 shows the shape of the principal parametric resonance in the $(\delta\omega, \mu)$-plane. Other resonance tongues exist, associated with different rational ratios between the excitation frequency and the natural frequency, but they are harder to excite.

In any case the fact that an external signal of pulsation 2ω can be used to generate oscillations of pulsation ω is a signature of the intrinsic nonlinearity of the parametric instability, which may not be obvious from the fact that Eq. (4.40) is seemingly linear in x. The clue of the paradox lies in the fact that the external forcing makes the system nonautonomous and thus requires the introduction of a new state variable, the forcing phase $\varphi = 2\omega t$ (see Section 1.1). Therefore Eq. (4.40) features the term $\cos\varphi \times x$, which makes it nonlinear.

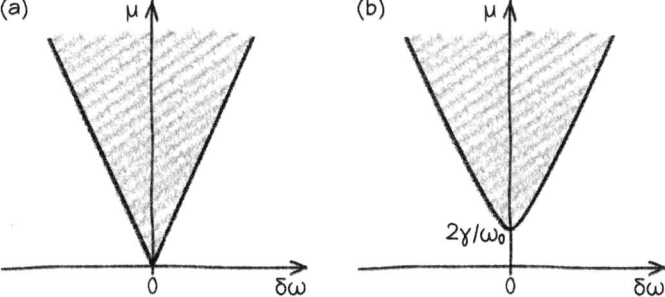

Figure 4.17: Region of the $(\delta\omega, \mu)$-plane where parametric oscillations occur: (a) $\gamma = 0$, (b) $\gamma > 0$.

4.3.2 Synchronization of oscillators: Arnold tongues

In the previous sections, we have studied resonances in damped nonlinear passive oscillators. The general question we will address now is: What does happen when a self-sustained oscillator of period T interacts with another oscillator with a different period? For simplicity, we will consider only the simpler case of an external driving cycle of fixed period T_0, i. e., when the interaction is unidirectional.

This case is relevant in many different situations encountered in physical, chemical, biological systems, etc. One interesting example is that of our biological clocks being forced by the day/night cycle and synchronizing to it. Yet, much of our results presented in this section actually apply to the more complex case of bidirectional interaction. Moreover, we will only focus on how the period T of the oscillator is affected, leaving its amplitude aside. This considerably simplifies the treatment of the problem, allowing us to keep only one dynamical variable.

4.3.2.a Basic phase-locking

While the problem can be very complicated, formulating it in terms of dynamical system concepts uncovers most of the phenomenology. The starting point is that any oscillation follows a limit cycle in the phase space, which is a one-dimensional closed object homeomorphic to the unit circle S^1. Thus the oscillation state of the driven system can be described by single periodic variable, a *phase*, which we will denote φ. Depending on what we study, φ can be considered as visiting the entire real axis or restricted to the interval $[0, 2\pi[$, provided that we remember that all the values $\varphi + 2n\pi$ are associated with the same point on the limit cycle.

Since there is no need to follow the phase φ in continuous time, it is sufficient to perform a *stroboscopic sampling*: at the beginning of each driving period (thus at times nT_0), we ask what precisely is φ_n, the current value of the oscillator phase φ. This in fact a special type of Poincaré section, associated with the crossing of a surface of constant phase of the driving. Because of determinism, the phase φ_n must be a function $\varphi_n =$

$F(\varphi_{n-1})$ of the phase at the previous sampling time, since φ_{n-1} completely describes the initial condition.

Let us first consider the case where the driving strength is zero. Then the phase dynamics only results from the period mismatch between the oscillator and the external cycle. Since $\varphi(t) = 2\pi t/T$, where t is time, and T is the oscillation period of the driven oscillator, we have that

$$\varphi_{n+1} \equiv \varphi((n+1)T_0) = \varphi(nT_0) + 2\pi\frac{T_0}{T} = \varphi_n + 2\pi\frac{T_0}{T},$$

which can be rewritten as $\varphi_{n+1} = \varphi_n + \Delta$ with $\Delta = 2\pi\frac{T_0}{T}$. Since phases are defined up to a shift of 2π, the phase shift Δ may be chosen so that $\Delta \in [0, 2\pi[$ or $\Delta \in [-\pi, \pi[$, but it is common to keep its original value to keep track of the true ratio of the original periods, for example, in the case of subharmonic forcing.

The dynamics of $\varphi_{n+1} = \varphi_n + \Delta$ depends crucially on whether $\Delta/2\pi$ is an irrational or rational number. At this stage, this is only related to the relative values of the two periods and does not result from an interaction. Nevertheless, these are the two behaviors that we will encounter in the general case.

When $\Delta = 2\pi\frac{p}{q}$, so that $\varphi_{n+q} = \varphi_n + 2p\pi = \varphi_n$, the motion is periodic with period qT_0, corresponding to the case where the forced oscillator undergoes exactly p periods during q cycles of the driving. This phenomenon is called $p : q$ phase-locking, since a permanent phase relation is observed between the oscillator and external driving. When $\Delta/2\pi$ is irrational, φ visits the entire interval $[0, 2\pi[$ without returning to its initial condition: we then have a *quasi-periodic* regime.

Let us now consider a nonzero coupling so that the evolution of the phase over one driving period is now influenced by the forcing. This effect is deterministic and can only depend on the initial oscillator phase, so that there must exist a function $V(\varphi)$ such that

$$\varphi_{n+1} = \varphi_n + \Delta + V(\varphi_n) \equiv F(\varphi_n). \tag{4.43}$$

The $V(\varphi_n)$ is often termed the *phase response curve*, as it describes the resetting experienced in response to the external forcing. Of course, it should be periodic in φ.

The simpler type of phase locking is when the oscillator recovers the same phase at each forcing period ($\varphi_{n+1} = \varphi_n$), prompting us to search for fixed points of mapping (4.43) (see Fig. 4.18a). We easily see that these fixed points can be obtained at the intersections of the curve $V(\phi)$ with a horizontal line of ordinate $-\Delta$ (Fig. 4.18b). Hence we see that a solution will exist as long as $\Delta \in [-V_{max}, -V_{min}]$, where $V_{min,max}$ are the minimal and maximal values taken by the function $V(\varphi)$. This indicates that the stronger the forcing, the more easily it will be able to compensate for large period mismatches.

To determine the stability of a fixed point satisfying $-\Delta = V(\varphi^*)$, we consider the evolution of a perturbation $\delta\varphi_n = \varphi_n - \varphi^*$. Following a reasoning already used previously, we obtain that

(a)

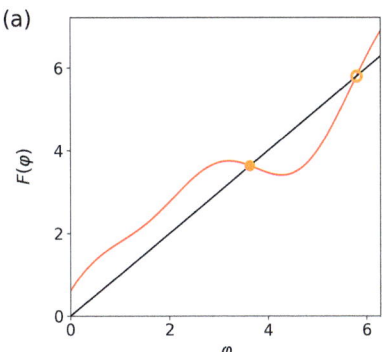

(b)

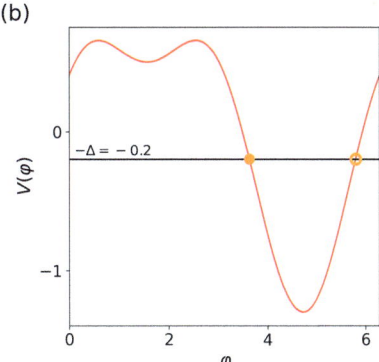

Figure 4.18: Graphical resolution of fixed points $\varphi_{n+1} = \varphi_n$ of Eq. (4.43). (a) Using the first return map: fixed points correspond to the intersections of the curve $y = F(\phi)$ with the line $y = x$. The intersection for which $|F'(\phi)| < 1$ is stable (orange point), and the one with $|F'(\phi)| > 1$ is unstable (orange circle). (b) Alternative solving method by searching the intersections of $V(\phi)$ with the line $y = -\Delta$. As discussed in the main text, the intersection corresponding to a negative slope for V' is stable (provided that $-2 < V'(\varphi^*) < 0$), and that with a positive one is unstable.

$$\delta\varphi_{n+1} = \delta\varphi_n + V'(\varphi^*)\delta\varphi_n = (1 + V'(\varphi^*))\delta\varphi_n. \tag{4.44}$$

Hence a fixed point will be stable if and only if $-2 < V'(\varphi^*) < 0$ (corresponding to $-1 < F'(\varphi^*) < 1$), indicating that only a fixed point with negative $V'(\varphi^*)$ can be stable. This makes sense, since a positive fluctuation of the phase must be compensated by a greater negative dephasing. Capitalizing on our analysis of the bifurcations of periodic orbits in Section 4.2.2, we note that the case $V'(\varphi^*) = 0$ ($F'(\varphi^*) = 1$) corresponds to a saddle-node bifurcation, where two new periodic orbits appear, one stable and the other unstable, which thus constitutes the mechanism by which phase locking occurs.

Thus a certain amount of forcing is needed to achieve the synchronization of the oscillator to the external signal. However, we see that if the forcing is too strong, then the phase-locked orbit becomes unstable when $V'(\varphi^*) = -2$ ($F'(\varphi^*) = -1$). The fact that an eigenvalue crosses -1 indicates us that this corresponds to a period-doubling bifurcation, where the same oscillator phase is recovered every other forcing period (see Section 4.2.2.b).

Hence the external forcing should be strong enough to lock the oscillator period to the driving period, but not too strong, to avoid destabilizing the phase-locked regime toward a more complex regime. This is important in the case of the so-called "circadian" clocks that reflect the astronomical time inside our body and orchestrate our physiology across the day–night cycle (Pfeuty et al., 2011). A period doubling would imply a change in behavior from one day to the next one.

This simple view is rigorous for weak forcing, where $V(\varphi)$ and Δ are small. When forcing is stronger, the analysis should take into account that all phases and phase shifts are considered modulo 2π, whereas the total phase shift during one cycle may actually be larger than 2π. Nevertheless, the general conclusion that a larger period mismatch between the driven and driving oscillators requires a stronger action of the external forcing for synchronization to take place remains mostly valid.

4.3.2.b General phase-locking and Arnold tongues

To investigate the extraordinary complexity of the general phase-locking problem, a much studied system is the famous circle map considered by Vladimir Arnold (1937–2010) (Arnold, 1965), where a cyclic variable $\theta \in [0,1[$ for simplicity, and the phase response curve is assumed to be sinusoidal:

$$\theta_{n+1} = \theta_n + \Omega + \frac{K}{2\pi}\sin(2\pi\theta_n) = F(\theta_n). \tag{4.45}$$

To study all the possible behaviors, it is generally sufficient to restrict Ω to a unit interval ($\Omega \in [-\frac{1}{2}, \frac{1}{2}[$ or $\Omega \in [0,1[$), given the periodicity in θ, which means that the cases Ω and $\Omega + 1$ cannot be distinguished under the restriction $\theta \in [0,1[$. Higher values of Ω can be understood as describing the case where driving occurs at a lower period than the oscillator period but does not lead to new behavior. The constant $K \geq 0$ quantifies the strength of the driving. Importantly, the map (4.45) is monotonic and thus invertible for $K \leq 1$, as the slope of F varies between $1 + K$ and $1 - K$. More complex phenomena, such as bistability and chaos, appear for $K \geq 1$ (Fig. 4.19(a)).

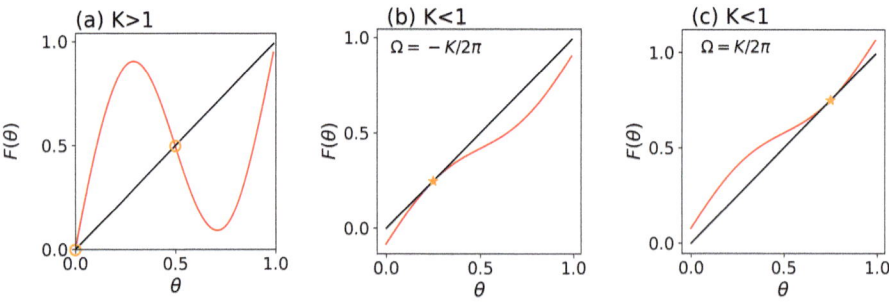

Figure 4.19: Arnold circle map. (a) Noninvertible case ($K > 1$). (b) and (c) Invertible case (here $K = 0.5$) for $\Omega = \pm\frac{K}{2\pi}$, where the graph of the circle map is tangent to the diagonal, indicating a saddle-node bifurcation. Circle maps for any value of Ω between those two limit cases have two intersections with the diagonal $y = x$, one stable and one unstable. The stable solution corresponds to a phase locking of the forced oscillator with the external driving.

Let us return to our analysis of 1:1 phase locking, which corresponds to the case where the oscillator and forcing periods are similar. Taking into account the periodicity in θ, this corresponds to $\Omega \sim 0$. We easily find that if $\Omega \in]-\frac{K}{2\pi}, \frac{K}{2\pi}[$, then there are exactly two fixed points $\theta_{s,u}^*$ for (4.45). The one where $K\sin(2\pi\theta^*)$ has a negative slope and is stable since mapping (4.45) has then a slope of absolute value smaller than 1.

Figures 4.19b and c show the limit cases $\Omega = \pm\frac{K}{2\pi}$, which correspond to the boundaries of the phase-locked regions. We can deduce from those graphs that the couple of fixed points $\theta_{s,u}^*$ appears (resp., disappears) through saddle-node bifurcations. Some authors consider $\Omega \in [0,1[; 1:1$ phase locking then occurs when $\Omega \in [0,\frac{K}{2\pi}] \cup [1-\frac{K}{2\pi},1[$.

The Arnold circle map (4.45) allows us to study more complex regimes of phase locking, where the original phase is recovered after q iterations: $\varphi_{n+q} = \varphi_n$. A useful quantity to characterize phase locking in this case is the *winding number*

$$w = \lim_{n\to\infty} \frac{\sum_{i=0}^n (\theta_{i+1} - \theta_i)}{n} = \lim_{n\to\infty} \frac{\theta_n}{n}, \qquad (4.46)$$

where the second expression holds if we let θ increase to infinity without restricting it to the unit interval, computing the difference $(\theta_{i+1} - \theta_i)$ literally from (4.45). The winding number expresses the average phase drift per cycle of the oscillator compared to the external forcing.

Because the winding number w only depends on the asymptotic dynamics, whether it is a rational or irrational number indicates the nature of the dynamical regime. When $w = \frac{p}{q}$ with $p, q \in \mathbb{N}$, we will have asymptotically that $\theta_{n+q} = \theta_n + p$, corresponding to a phase-locked regime due to periodicity. When w is instead irrational, all phase differences are observed, corresponding to quasi-periodicity. It can be shown that if the circle map is a homeomorphism (hence for $K < 1$ here), then the winding number is unique and does not depend on the initial condition (Katok and Hasselblatt, 1995). When the circle map is not a homeomorphism (hence for $K > 1$), a winding number can be defined for a given initial condition but is not unique.

In fact, Ω represents the winding number at zero forcing ($K = 0$). Since rational numbers are dense in the unit interval but occupy a set of measure 0, we then observe quasi-periodicity with probability 1. When the forcing strength K increases, phase locking associated with rational winding number occurs in finite intervals that grow with K. For instance, period-1 phase locking occurs for $\Omega \in [0,\frac{K}{2\pi}]$ (with winding number $w = 0$) or $\Omega \in [1-\frac{K}{2\pi},1]$ (with winding number $w = 1$). Hence a phase-locked regime becomes more and more likely as K increases.

In Fig. 4.20, we show the regions of the (Ω, K)-plane where stable phase-locked regimes exist associated with a winding number given by a simple rational fraction (here we restricted ourselves to denominators of 7 or lower). We see that each phase-locking region of fixed winding number extends as K increases, forming what is called an Arnold tongue. We also note that when the denominator of the fraction is larger, the tongue is much narrower. In fact, it is known that the width $\Delta\Omega$ of a tongue of ratio p/q scales as k^q when $k \to 0$ and as q^{-3} at finite k (Ecke et al., 1989).

Although the tongues drawn in Fig. 4.20 occupy a moderate percentage of the $[0,1]$ interval at $K = 1$, they are only a finite subset of an infinity of tongues associated with ratios of arbitrarily high denominators. It turns out that for $K = 1$, almost all values of Ω lead to a phase-locked regime (i. e., with probability 1). There are still quasi-periodic

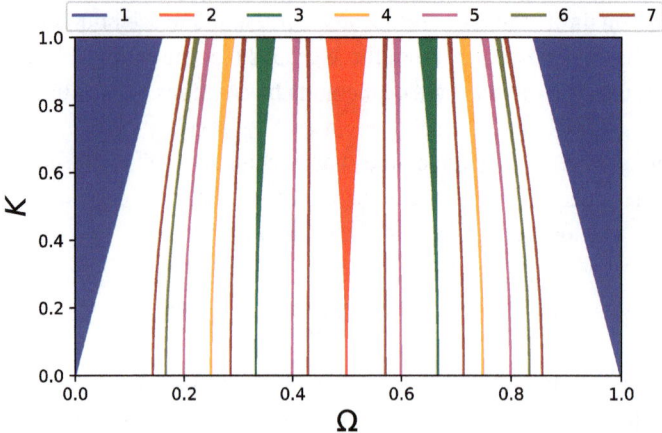

Figure 4.20: Arnold tongues showing the phase-locking regions corresponding to the simplest fractions of Eq. (4.46).

regimes separating two phase-locked regimes, but they form a *fractal* set (Jensen et al., 1984), a concept that will be explored in Section 6.1.2.

To better appreciate the complexity of the alternation of quasi-periodicity and phase-locking at $K = 1$, Fig. 4.21 displays the variation of the winding number w as a function of Ω for $K = 1$. Each horizontal plateau, where the winding number is rational and constant, corresponds to one phase-locking regime, and there is one plateau for each rational $\frac{p}{q}$. The complexity of this curve is such that it is known as the Devil's staircase.

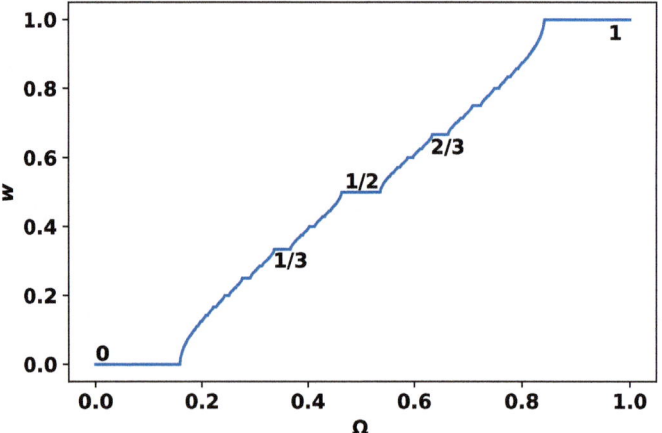

Figure 4.21: The Devil's staircase: the winding number defined by (4.46) is plotted as a function of Ω. The winding number of the most prominent plateaus is indicated. There is an infinity of points where a quasi-periodic regime is observed but they make up a set of measure 0.

The complex organization of Arnold tongues is reflected in the so-called Farey tree based on Farey arithmetic: between the tongues associated with rational number $\frac{p}{q}$ and $\frac{p'}{q'}$, the most prominent Arnold tongue is associated with rational number $\frac{p+p'}{q+q'}$ ("Farey sum"). Starting with the two "mother tongues" associated with the ratios $0/1$ and $1/1$, we can thus derive the so-called Farey tree by inserting at each step between two adjacent tongues the tongue associated with the Farey sum of the two ratios. The first stages of this construction are shown in Table 4.1.

Table 4.1: Farey tree showing the hierarchical structure of the Arnold tongues.

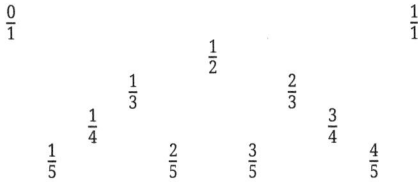

4.4 Conclusions

In this chapter, we have studied how the bifurcation of a fixed point through a Hopf bifurcation leads to the emergence of a limit cycle, the type of attractor associated with oscillatory asymptotic behavior. According to the Poincaré–Bendixson theorem, limit cycles and fixed points are the only generic asymptotic behaviors in a 2D phase space. We then showed how to determine the stability of a periodic orbit by studying the stability of fixed points of a Poincaré return map and discussed the different bifurcations that periodic orbits can experience. Finally, we studied different cases of driven oscillations and in particular the synchronization of a self-sustained oscillator with an external driving cycle, leading to either periodic or quasi-periodic behavior. In Chapter 5, we will see that these regimes correspond to a dynamics on an invariant torus in a three-dimensional phase space.

Exercises

Model of cell division

The cell division cycle is mainly controlled by two interacting proteins: the cyclin-dependent kinase (cdc2) and the cyclin. Those two proteins can combine to form a complex, named MPF (for maturation promoting factor). The dynamics of the concentration of this complex regulates cell division. The goal of the exercise is to study a model of the formation and activation of MPF that displays spontaneous oscillations and thus can predict cycles of successive cellular division. This exercise is based on (Tyson, 1991).

The interplay between cyclin and cdc2 follows a complex cycle, which implies numerous steps (synthesization and degradation of cyclin from amino acids, phosphorylation and dephosphorylation of the different proteins, MPF formation and activation). The kinetic equations describing this complex cycle can be simplified. In particular, as some variables evolve on a very short time scale compared to others, their dynamics are decoupled from the rest of the system. Finally, the dynamics can be reduced to the study of the following system:

$$\frac{du}{dt} = k_4(v - u)(a + u^2) - k_6 u, \tag{4.47a}$$

$$\frac{dv}{dt} = k_1 \frac{[aa]}{[CT]} - k_6 u, \tag{4.47b}$$

where u is the fraction of activated MPF in the cell, and v is the total fraction of cyclin in its different forms. The parameters are linked to the kinetic constants of the reactions as well as the concentrations of the different components. All the parameters are positive, and denoting $\beta = k_6/k_4$, their values of interest obey to the inequality $a \ll \beta \ll 1$. In the following, we will typically consider $\beta \simeq 10^{-2}$ and $a \simeq 10^{-4}$.

1. The nullcline deduced from Eq. (4.47a) corresponds to a function $v = f(u)$. Show that this function has a maximum for $(u, v) \simeq (\sqrt{a}, \beta/(2\sqrt{a}))$ and a minimum for $(u, v) \simeq (\sqrt{\beta}, 2\sqrt{\beta})$.

2. The position of the v-nullcline depends on the exact values of the parameters. Draw the three possible relative positions of the two nullclines. How many fixed points the system has?

3. Compute the Jacobian matrix J of the dynamical system. Show that the real part of the two eigenvalues λ_1 and λ_2 of this matrix have always the same sign.

4. We denote by $T(u, v)$ the trace of J. Show that the determination of the stability of the fixed point consists in determining the sign of the function $T(u, f(u))$.

5. Show that $T(u, f(u)) = -k_4(a + u^2)f'(u)$. Deduce a graphical condition for the fixed point to be stable (respectively, unstable).

6. Cell division is controlled by the activity of MPF whose concentration peaks during *metaphase* (one of the phases of the cell division). Show that the model presented in this exercise predicts cell division cycles for some values of the parameters. Draw in the phase space the corresponding nullclines and vector field. Those cycles could correspond to the rapid division cycles that occur for early embryo.

7. The typical values of the parameter are

$$k_1 \frac{[aa]}{[CT]} = 0.015 \, \text{min}^{-1}, \quad k_4 = 10 \rightarrow 1000 \, \text{min}^{-1},$$

$$k_6 = 0.1 \rightarrow 10 \, \text{min}^{-1}, \quad a = \frac{k_4'}{k_4} \quad \text{with } 0.018 \, \text{min}^{-1}.$$

Numerically integrate the system of Eqs. (4.47) and plot typical trajectories in the phase space together with the nullclines. Check that periodic solutions are indeed obtained for appropriate parameter values. Plot the concentration dynamics as a function of time to observe the spiking behavior of the MPF concentration.

Bullard dynamo and Faraday disk

This exercise is a continuation of one of the exercises of Chapter 3 and is based on (Laroche et al., 2012). A Bullard dynamo is a mechanical system that produces spontaneously a current I in a coil when a conducting disk is rotated at a high enough rotation rate Ω_1 (see exercise *Bullard dynamo* in Chapter 3 and in particular Figure 3.18).

In the present exercise, we use the current I generated by a dynamo to feed a Faraday disk. The schematic of the circuit is shown in Fig. 4.22. The Faraday disk is submitted to an external magnetic field $\mathbf{B}_0$. The two disks are identical, their radius is a, and their inertial moment is J. They also experience the same friction phenomenon characterized by a coefficient λ.

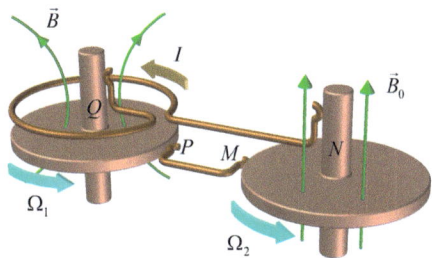

Figure 4.22: Reprinted from (Laroche et al., 2012) with the permission of AIP Publishing. Sketch of a Faraday disk coupled to a Bullard dynamo.

In this system the parameters that the operator can tuned are the torque Γ imposed to the dynamo and the magnetic field $\mathbf{B}_0$ imposed to the Faraday disk. Those two parameters are adjustable. The variables describing the dynamics of the system are the current I circulating in the coil and the angular rotation rates of each of the conducting disk: Ω_1 for the dynamo and Ω_2 for the Faraday disk.

The equation governing the dynamics of the current is

$$L\frac{dI}{dt} + RI = M\Omega_1 I - \alpha B_0 \Omega_2, \tag{4.48}$$

where R and L are respectively the resistance and inductance of the circuit, and M is the mutual inductance between the coil and the conducting disk of the dynamo. The term $-\alpha B_0 \Omega_2$ is the electromotive force of the Faraday disk with $\alpha = a^2/2$.

The equation governing the rotation rate of the Bullard dynamo is

$$J\frac{d\Omega_1}{dt} = \Gamma - \lambda\Omega_1 - MI^2,$$ (4.49)

and the one governing the rotation rate of the Faraday disc is

$$J\frac{d\Omega_2}{dt} = \alpha B_0 I - \lambda\Omega_2.$$ (4.50)

1. Show that this system has two fixed points:
   - a trivial one ($I = 0, \Omega_1 = \frac{\Gamma}{\lambda}, \Omega_2 = 0$);
   - two nontrivial ones, ($I^*, \Omega_1^* = \frac{1}{\lambda}(\frac{(\alpha B_0)^2}{M} + \Gamma_0), \Omega_2^* = \frac{\alpha B_0 I^*}{\lambda}$) with $\Gamma_0 = \frac{\lambda R}{M}$ and $(I^*)^2 = \frac{1}{M}(\Gamma - \Gamma_0 - \frac{(\alpha B_0)^2}{M})$.
2. Compute the Jacobian matrix of the system.

First, we study the stability of the trivial fixed point ($I = 0, \Omega_1 = \frac{\Gamma}{\lambda}, \Omega_2 = 0$).
3. Show that the characteristic polynomial of the Jacobian matrix at this point is

$$p(X) = -\left(X + \frac{\lambda}{J}\right)\left[X^2 + \left(\frac{\lambda}{J} + \frac{M(\Gamma_0 - \Gamma)}{\lambda L}\right)X \right.$$
$$\left. + \frac{\alpha^2 B_0^2 + M(\Gamma_0 - \Gamma)}{LJ}\right].$$ (4.51)

4. If $\Gamma \ll \Gamma_0$, what is the stability of the fixed point?
5. Discuss the loss of stability of the fixed point in the two following cases:
   (a) ($\frac{\lambda}{J} + \frac{M(\Gamma_0-\Gamma)}{\lambda L}$) becomes negative while $\alpha^2 B_0^2 + M(\Gamma_0 - \Gamma)$ stays positive;
   (b) $\alpha^2 B_0^2 + M(\Gamma_0 - \Gamma)$ becomes negative while ($\frac{\lambda}{J} + \frac{M(\Gamma_0-\Gamma)}{\lambda L}$) stays positive.
6. Which of these two ways of loosing stability can give rise to an oscillating solution? What is the criterion on B_0 to observe it?

We now study the nontrivial fixed point.
7. Show that the characteristic polynomial of the Jacobian matrix at this point is

$$p(X) = -\left(X + \frac{\lambda}{J}\right)\left[X^2 + \left(\frac{\lambda}{J} - \frac{\alpha^2 B_0^2}{\lambda L}\right)X + \frac{2(MI^*)^2}{LJ}\right].$$ (4.52)

8. Discuss the stability of the nontrivial fixed point.
9. Draw a diagram in the parameter space (B_0, Γ) showing the domains of stability of the different solutions.

Property of the slopes of the nullclines

We consider the bidimensional system

$$\frac{dx}{dt} = f(x,y),$$
$$\frac{dy}{dt} = g(x,y).$$

1. Show that the slopes p_x (respectively, p_y) of the x-nullcline (resp., y-nullcline) are given by

$$p_x = -\frac{\partial f/\partial x}{\partial f/\partial y},$$
$$p_y = -\frac{\partial g/\partial x}{\partial g/\partial y}.$$

2. Show that in the vicinity of a Hopf bifurcation,

$$\frac{\partial f}{\partial x}\frac{\partial g}{\partial y} - \frac{\partial g}{\partial x}\frac{\partial f}{\partial y} > 0,$$
$$\frac{\partial f}{\partial x} + \frac{\partial g}{\partial y} = \epsilon$$

with $|\epsilon| \ll 1$. Discuss the sign of ϵ at each side of the bifurcation.

3. Deduce that $\frac{\partial f}{\partial x}$ and $\frac{\partial g}{\partial y}$ are of opposite signs and that the same is true for $\frac{\partial g}{\partial x}$ and $\frac{\partial f}{\partial y}$.

4. Show that in the vicinity of a Hopf bifurcation the slopes of the nullclines are necessarily of the same sign.

The Sel'kov glycolysis model

In 1968, the russian biophysicist E. E. Sel'kov wrote a mathematical model for the glycolysis process, where glucose is broken into smaller molecules to produce ATP (Sel'Kov, 1968). In some limit, this model can be reduced to the following equations:

$$\dot{x} = f(x,y) = -x + ay + x^2 y = -x + h(x,y), \tag{4.53a}$$
$$\dot{y} = g(x,y) = b - ay - x^2 y = b - h(x,y), \tag{4.53b}$$

where $a, b \in [0,1]$.

1. Determine the fixed point of the system.
2. Draw the nullclines of the system in the phase space.
3. Show that the fixed point loses its stability for

$$a_c(b) = \frac{-(1+2b)^2 + \sqrt{1+8b^2}}{2}.$$

4. Integrate numerically the system for $b = 0.6$ and different values of a below and above a_c. Verify that the asymptotic behavior when $a > a_c$ is periodic.

5. Draw the stability diagram of the Sel'kov model, showing the domains of the (a, b)-plane corresponding to identical asymptotic behaviors.

5 Quasi-periodicity and strange attractors

In the previous chapters, we have seen that when the dimension of the phase space increases, the complexity of the possible asymptotic behaviors also increases: in 1D, the only possible asymptotic behavior is a stationary dynamics, whereas in 2D, oscillations can occur. In 3D and above, deterministic systems can display aperiodic asymptotic behaviors, as we will see in the present chapter.

In practice, complex dynamical behaviors persisting over arbitrarily long times present new challenges for us as they may have different origins, which we need to discriminate to ascertain the nature of the dynamics. The first question when faced with an aperiodic signal is therefore whether the signal is generated by an essentially random and stochastic process or arises from deterministic equations of motion. Even when the dynamics is deterministic, as in this textbook, quasi-periodicity and chaos have quite different properties. This implies that we must not only understand how each type of motion is associated with a specific type of attractor, but also how can we characterize the latter to understand the underlying dynamical mechanisms.

5.1 Introduction

Two types of attractors generated by a deterministic dynamics can only be observed when the dimension of the phase space is at least three. When the trajectory in phase space wraps up around an invariant torus (e. g., Fig. 5.1(a)), the dynamics is quasi-periodic, a behavior that we already encountered in Section 4.3.2. Strange attractors such as that shown in Fig. 5.1(c) are generated by chaos, a dynamical regime that we will characterize by its extreme sensitivity to initial conditions.

The two types of regimes generate complicated aperiodic signals (Fig. 5.1(b,d)) that are difficult to discriminate based on their appearance but have very different levels of complexity, as we will see. To design Fig. 5.1, we used the following three-dimensional models that serve as examples of the two dynamical regimes:

- The driven van der Pol oscillator is used to illustrate quasi-periodic behavior (Fig. 5.1(a,b)):

$$\ddot{x} + \varepsilon(x^2 - 1)\dot{x} + x = \Gamma \cos \omega t.$$

This is a self-sustained oscillator driven by an external frequency, a case that we studied in Section 4.3.2. As we have learned in Section 1.1.2, this system can be rewritten as the following first-order autonomous system:

$$\begin{cases} \dot{X}_1 = X_2, \\ \dot{X}_2 = -\varepsilon(X_1^2 - 1)X_2 - X_1 + \Gamma \cos X_3, \\ \dot{X}_3 = \omega. \end{cases} \tag{5.1}$$

https://doi.org/10.1515/9783110677874-005

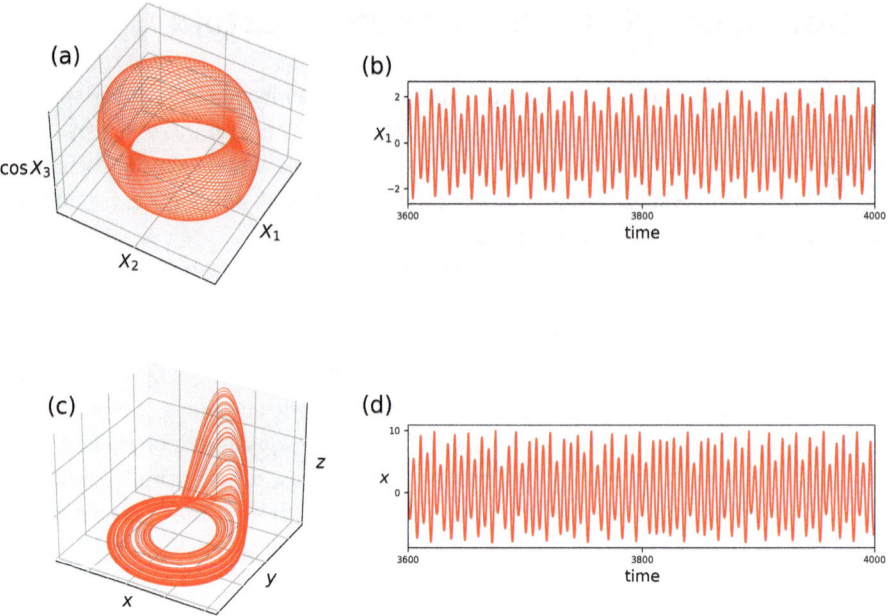

Figure 5.1: (a) Attractor of the forced Van der Pol oscillator of Eq. (5.1) with $\varepsilon = 0.125$, $\omega = 2/(1 + \sqrt{5})$, and $\Gamma = 0.4$; (b) corresponding asymptotic time series of the variable X_1; (c) attractor of the Rössler system of Eq. (5.2) with $a = b = 0.2$ and $c = 5$; (d) corresponding asymptotic time series of the variable x.

– Figs. 5.1(c) and (d) were computed with the Rössler system

$$\begin{cases} \dot{x} = -y - z, \\ \dot{y} = x + ay, \\ \dot{z} = b + z(x - c), \end{cases} \qquad (5.2)$$

which will be our benchmark for studying chaotic behavior.

5.2 Periodic and quasi-periodic motion on a torus

5.2.1 Topology of the torus

In Chapter 4, we learned that when a periodic behavior emerges, its attractor is topologically equivalent to the unit circle S^1. This allows oscillations to take place in the phase plane, although they may sometimes require three dimensions to unfold. When we couple two oscillators, their combined state space has the topology of a product space, namely the two-dimensional torus $T^2 = S^1 \times S^1$ (Fig. 5.2). This object can only be embedded in a three-dimensional phase space but is naturally charted using two phases (ϕ_1, ϕ_2) as coordinates, describing the rotations associated with the two oscillators.

The first rotation is along the large circle, which is called the *toroidal direction*, and unfolds at frequency f_1. The other rotation is along the small circle, which is called the *poloidal direction* and is characterized by the frequency f_2 (Fig. 5.2).

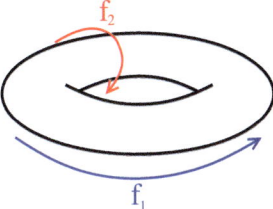

Figure 5.2: Schematic representation of an invariant torus in the phase space.

From Section 4.3.2, where we studied the dynamics of coupled oscillators using circle maps, we know that the torus is not always completely visited. Indeed, we found that two behaviors are possible: periodic, when the oscillators are phase-locked, the ratio of their frequencies being a rational number, and quasi-periodic, when this ratio is irrational. Here we revisit this question by considering how it is organized in the phase space.

To understand how the trajectories are organized, it is convenient to unfold the torus by cutting it along the poloidal and toroidal circles. This allows us to represent the dynamics in the plane (ϕ_1, ϕ_2), more precisely, in the domain $[0, 2\pi] \times [0, 2\pi]$, and to better visualize it (Fig. 5.3).

Trajectories confined in the flattened torus (Fig. 5.3a) are constrained by two properties: (1) the boundary conditions are periodic, and (2) they cannot intersect. These constraints forbid two trajectories to diverge from each other, greatly restricting their global

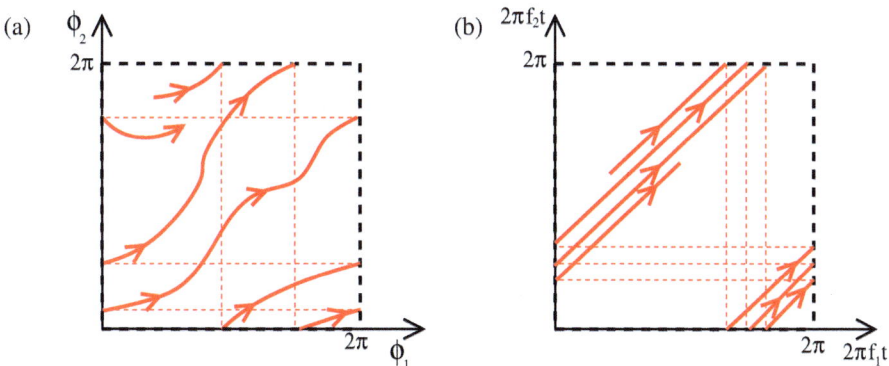

Figure 5.3: Trajectories on the flattened torus. (a) The constraints of periodic boundary conditions and nonintersection forbid any divergence of the trajectories. (b) There is a change of coordinates which transforms the trajectories drawn in (a) into parallel lines traveling at constant velocity.

organization. Close trajectories must remain close to each other, without significantly changing their global orientation because of the periodic boundary conditions. Looking at Fig. 5.3, we can convince ourselves that it is in fact always possible to find a change of variables so that the trajectories in the square are straight lines parallel to each other (Fig. 5.3b) traveling at constant speed. We then have $\phi_i = 2\pi f_i t$, and the common slope of the lines $\frac{d\phi_2}{d\phi_1}$ is equal to the frequency ratio f_2/f_1.

A system of parametric equations that describe the dynamics is

$$X_1(t) = r\cos(\omega_1 t) + \cos(\omega_2 t)\cos(\omega_1 t),$$
$$X_2(t) = r\sin(\omega_1 t) + \cos(\omega_2 t)\sin(\omega_1 t), \qquad (5.3)$$
$$X_3(t) = \sin(\omega_2 t),$$

where $\omega_i = 2\pi f_i t$.

5.2.2 Periodic behavior

If the ratio $\frac{f_1}{f_2}$ is a rational number $\frac{p}{q}$ (with relatively prime $p, q \in \mathbb{N}$), then the trajectory closes after exactly p revolutions along the toroidal direction, which take the same time as q revolutions along the poloidal direction, since we have $pT_1 = qT_2 = T$ where $T_i = 1/f_i$ are the two periods. The dynamics is periodic with period T. In the three-dimensional phase space, the trajectory comes back to its initial condition after a finite number of turns around the two circles (Fig. 5.4a).

Again, we obtain a simpler view by performing a Poincaré section, conveniently achieved with a plane. The periodic orbit has a finite number of intersections with the

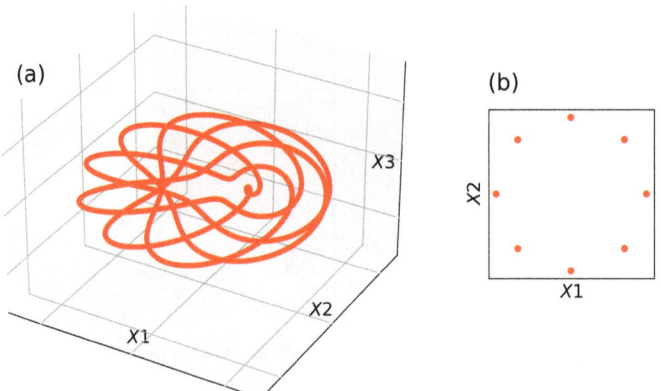

(a)

(b)

Figure 5.4: (a) Periodic behavior of system (5.3) corresponding to the ratio $f_1/f_2 = 5/8$. (b) Poincaré section using the plane $X_3 = 0$, $\dot{X}_3 > 0$, displaying 8 intersections or, equivalently, using a stroboscopic method at frequency f_2.

section plane (Fig. 5.4b). If $\mathcal{P}$ denotes the map of first return in the Poincaré section, and $X_1, X_2, \ldots, X_q$ are the successive intersections with the plane, then we have

$$\mathcal{P}(X_i) = X_{i+1} \quad \text{and} \quad \mathcal{P}^q(X_i) = X_i.$$

The number of intersections depends on the orientation of the plane with respect to the two circles defining the torus. If it is aligned with the toroidal direction and transverse with the poloidal one (as in Fig. 5.4b), then each oscillation performed in time T_2 around the latter yields one intersection point. There will be thus q intersection points, since the period of motion is $T = qT_2$.

If the plane is aligned with the poloidal direction and transverse to the toroidal one, then the plane will be intersected p times before coming back to the initial condition, since the period $T = pT_1$, and we will have p intersections. Choosing one or the other orientation depends very much on which of the two coupled oscillations we want to consider as a reference.

5.2.3 Quasi-periodicity

When the ratio f_1/f_2 is irrational (f_1 and f_2 are then said to be *incommensurable*), the two oscillations have no common period. The trajectory never returns to a previously visited position (otherwise, it would be periodic) but fills densely the phase space: a trajectory starting anywhere will come arbitrarily close to any point on the torus surface, as Fig. 5.5(a) shows for the parametric system (5.3). Contrary to the periodic case, the invariant set of the flow in the phase space is now two-dimensional.

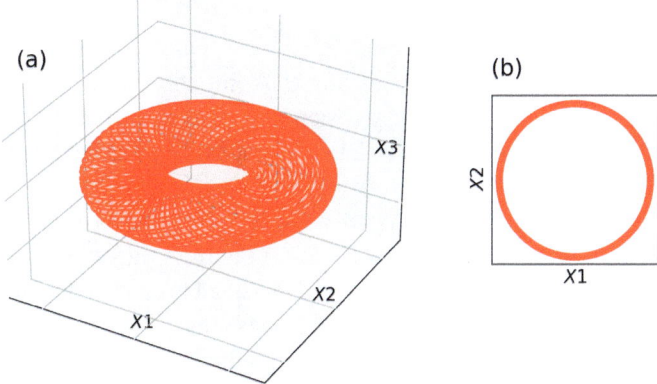

Figure 5.5: (a) Quasi-periodic behavior of system (5.3) on a torus corresponding to the ratio $\frac{f_1}{f_2} = \frac{1}{\phi}$, where $\phi = \frac{1+\sqrt{5}}{2}$ is known as the golden ratio. (b) Poincaré section using the plane $X_3 = 0$, $\dot{X}_3 > 0$ or, equivalently, using a stroboscopic method at frequency f_2.

In this case, the Poincaré section is a closed continuous curve C that fills the intersection of the torus with the section plane. This curve is invariant under the action of the Poincaré map, i. e., $\mathcal{P}(C) = C$.

The dynamics is stably attracted to the invariant torus, but what about the stability of the quasi-periodic trajectory with respect to perturbations inside the torus? To answer this question, we turn to the forced van der Pol system (5.1). Restricting ourselves to the Poincaré section, we analyze the dynamics along the invariant circle. Figure 5.6(a) shows a projection of the attractor of Fig. 5.1 in the (X_1, X_2) plane. A stroboscopic Poincaré section at the driving frequency is shown in Fig. 5.6(b). Since the section is homeomorphic to a circle, we can parameterize it with an angle variable, which we choose as discussed in the caption of Fig. 5.6(c). Associating a phase φ_n for each intersection with the Poincaré section, we can construct a first return map by plotting φ_{n+1} vs. φ_n of the points along the Poincaré section (Fig. 5.6(c)).

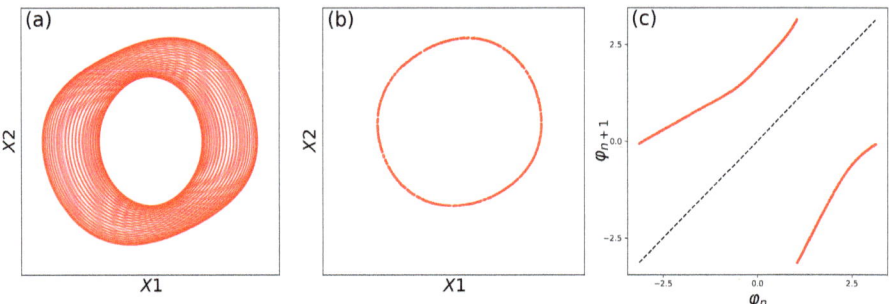

Figure 5.6: (a) Attractor of the driven Van der Pol oscillator of Eq. (5.1) with $\varepsilon = 0.125$, $\omega = 2/(1 + \sqrt{5})$, and $\Gamma = 0.4$ projected in the (X_1, X_2) plane. (b) Poincaré section using stroboscopy at frequency $\omega/2\pi$; (c) first return map obtained by plotting φ_{n+1} vs. φ_n, where φ is the angle of a polar coordinate system whose origin is fixed at the centroid of the circle. Note that the discontinuity of the curve is due to the 2π-periodicity of φ.

The first return map in Fig. 5.6(c) is monotonous and thus invertible. The interval $[-\pi, \pi]$ is mapped into itself once, so that the average slope is exactly one. Thus the distance between adjacent trajectories remains constant on average with transient fluctuations, and their order is preserved.

This insensivity to initial conditions is illustrated in Fig. 5.7, which shows three trajectories starting from close initial conditions. In Fig. 5.7(a) the time traces of X_1 for the three initial conditions are almost superimposed, and the small mismatch does not evolve. Accordingly, the corresponding trajectories in the phase space remain close to each other (Fig. 5.7(b)). These observations are consistent with the fact that trajectories cannot cross on the surface of an invariant torus.

To conclude, although quasi-periodic signals such as those plotted in Fig. 5.7(a) look complicated, they have no intrinsic complexity. They can be considered as compositions of elementary oscillations and actually can be generated by linear systems.

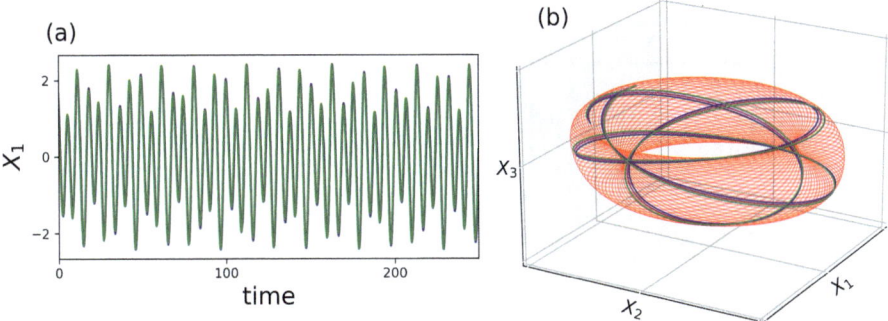

Figure 5.7: (a) Three time series of the X_1 variable of the driven van der Pol system defined by Eqs. (5.1) and starting from close initial conditions on the torus. (b) Corresponding trajectories in the phase space.

In this section, we considered the two-frequency dynamics associated with a T^2 torus, which exists in a phase space of dimension equal to or larger than 3. We can imagine higher-dimensional quasi-periodic systems resulting from the interaction of n frequencies and evolving on a T^n torus. It might then be difficult to picture the attractor in high-dimensional phase spaces, and in the limit of large n, the signal will be almost indistinguishable from pure noise.

Interestingly, the Russian physicist Lev Landau once hypothesized that turbulent motion could be explained in terms of high-dimensional quasi-periodic regimes. This picture was challenged by D. Ruelle and F. Takens in 1971, in the paper *"On the Nature of Turbulence"* (Ruelle and Takens, 1971). They showed that high-dimensional invariant tori are fragile, and that it is much more likely that a topologically distinct kind of attractor is observed. In opposition to invariant tori, which are regular surfaces, these attractors have particular complex geometry and structure, to the point that Ruelle and Takens called them *strange attractors*. Understanding the structure of these attractors is the subject of the next section.

5.3 Strange attractors and chaos

5.3.1 Attractors in phase space

Let us now consider the Rössler system (5.2) introduced in Section 5.1, which is also a three-dimensional nonlinear system displaying irregular dynamics for some parameter sets (Fig. 5.1(d)). If we compare the time evolution of the system starting from three close initial conditions on the attractor, then we observe that the three time series differ dramatically after some characteristic time (Fig. 5.8(a)) and that the corresponding trajectories in the phase space diverge quickly from each other (Fig. 5.8(b)). This behavior is utterly different from that of a quasi-periodic regime.

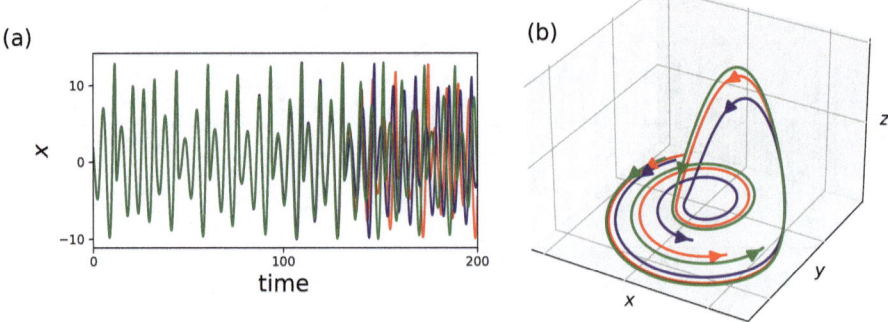

Figure 5.8: (a) Three time series of the X_1 variable of the Rössler system defined by Eqs. (5.2) for $(a, b, c) = (0.2, 0.2, 5.7)$, starting from close initial conditions on the attractor. (b) Trajectories in the phase space for $t \in [150; 160]$.

The sensitivity to initial conditions observed in Fig. 5.8 points to a much higher complexity in the phase space than for quasi-periodic regimes, where trajectories were restricted to the two-dimensional surface of a torus. Thus there must be specific mechanisms generating this "chaos," where trajectories separate exponentially fast. We expect that these mechanisms operate in the phase space where the complex organization of a strange attractor becomes apparent.

Figure 5.9 shows an attractor of the Rössler system, simultaneously illustrating its complexity and the clear existence of a sophisticated organization. We can see that tra-

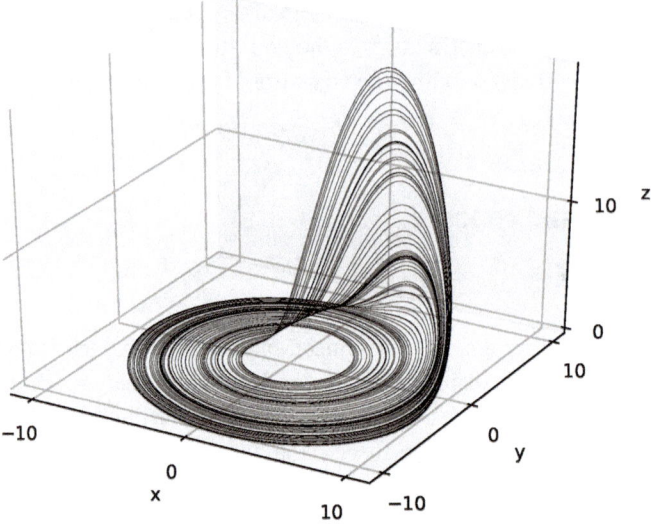

Figure 5.9: A long chaotic trajectory computed for $(a, b, c) = (0.2, 0.2, 5.7)$ explores the Rössler attractor and progressively covers it.

jectories close to the point $(0, 0, 0)$ spiral away from it in the (x, y)-plane, then take off abruptly in the z-direction when they are far enough from the origin, and follow a large reinjection loop that brings them back to the (x, y)-plane.

The alternation of the spiral around the origin and of the large reinjection loop is repeated indefinitely, but the system never returns exactly to a previously visited location. As a result, the trajectory of the system, which is clearly one-dimensional over a short time, gradually fills the attractor. This complex geometric object is close to being two-dimensional (Fig. 5.9) but is more than a surface, allowing a complex intertwining of the trajectories. As we will see, a strange attractor is a *fractal* object.

5.3.2 Characterizing determinism through the first return map

To simplify our representation of the dynamics of the Rössler system and gain insight into the chaos-generating mechanisms, we will now make use of a Poincaré section. Whereas the choice of a section surface is rather straightforward on a torus, it is more delicate for the Rössler attractor. We choose the $x = 0$, $\dot{x} > 0$ section plane because in this region, the trajectories are almost confined to the $z = 0$ plane (Fig. 5.10a) and the equations of motion are quasi-linear. Interestingly, the section of the attractor is then well approximated by a one-dimensional curve (Fig. 5.10b), which allows us to accurately locate points in the section with a single coordinate y.

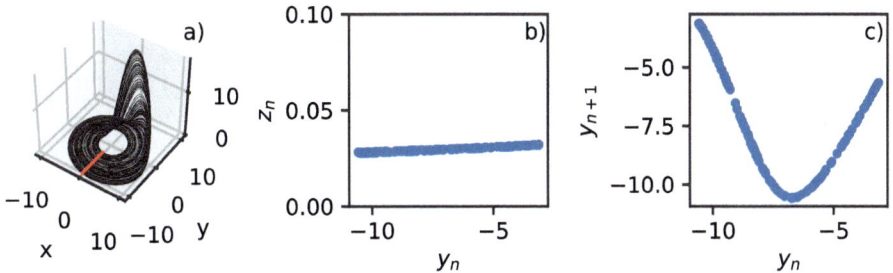

Figure 5.10: (a) Rössler attractor in phase space with the $x = 0$ Poincaré section plane; (b) Poincaré section of the attractor in the (y, z)-plane; (c) First return map, showing the coordinate y_{n+1} of the $(n + 1)$th intersection as a function of y_n.

If y_i denotes the position of the ith intersection with the section plane, then a good approximation of the map of first return in the section plane is plotting y_{n+1} vs. y_n. Indeed, we see in Fig. 5.10c that this plot is very close to the graph of a function $y_{n+1} = f(y_n)$ of the real line into itself. This demonstrates that the location of any point in the section can be (almost) predicted exactly from the location of the previous intersection. When such a simple return map can be obtained for a dynamical system, *it is a clear signature of the deterministic nature of the irregular dynamics observed*, which is especially useful to discriminate chaos from noise in experimental systems. It is not that uncommon to be able to construct such a one-dimensional return map, as strongly dissipative systems embedded in a three-dimensional phase space have typically almost 1D Poincaré section and thus an associated 1D return map.

What does the first return map in Fig. 5.10(c) tell us about the dynamics of the system? The first important feature is that the underlying continuous curve is not invertible, as two different points can have the same image. Thus there exist two well-separated parts of the Poincaré section, which are collapsed onto each other after completing a revolution on the attractor, forgetting their distinct past and sharing the same future. We may say that some states are *squeezed* onto each other.

This noninvertibility has a second consequence. By examining Fig. 5.10(c) we see that the V-shape of the graph implies that each subinterval where the map is monotonous is mapped to a larger interval, so that the average slope is larger than 1, which is confirmed by direct inspection. Since an error on y evolves as $\delta y_{n+1} = |f'(y_n)|\delta y_n$, this suggests that there is a number $\mu > 1$ such that $\delta y_n \sim \mu^n \delta y_0$. If small errors are thus growing exponentially, then this explains how trajectories can separate so fast from each other. We may say that different states are *stretched* apart.

These two processes are complementary: stretching separates neighboring trajectories, whereas squeezing maintains them in a bounded region.

5.3.3 Strange attractors are shaped by stretching, folding, and squeezing processes in phase space

5.3.3.a The recipe for generating a strange attractor

Let us now illustrate these two antagonistic processes more visually, again using Poincaré sections (Fig. 5.11). Since the overall motion of the Rössler system in the phase space largely consists of a rotation around a vertical axis going through the origin, we consider a series of Poincaré planes oriented vertically, all sharing this axis. As a trajectory rotates around the vertical axis, it will successively cross different planes. Examining how the intersection of the attractor with a plane is modified from a plane to the next one, we may hope to develop a vision of the geometric processes at play, as we show in Fig. 5.11.

In the first 6–7 plots of Fig. 5.11b, a stretching process is clearly visible as the length of the Poincaré section gradually increases. Then some sort of folding occurs in the remaining five planes, creating transiently a clear horseshoe shape. When the two branches of this horseshoe are squeezed onto each other, we almost recover the original shape, being ready for another turn.

Now we have to imagine that this sequence (stretch–fold–squeeze) is repeated indefinitely and that the strange attractor we observe in the phase space is invariant under this geometrical process. Thus what seems to be the same linear structure in the first and

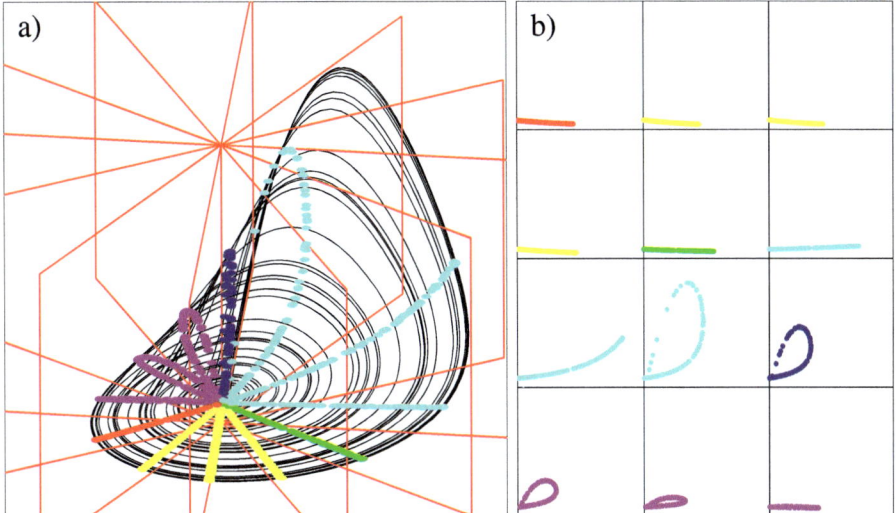

Figure 5.11: (a) Intersections of the Rössler attractor with several Poincaré planes sharing the same vertical axis going through the origin (here $a = 0.4287$, $b = 2.0$, and $c = 4.0$). (b) The different section planes are shown side by side, from top to bottom, and from left to right, allowing us to visualize how the Poincaré section of the attractor is modified as we wind around the axis.

last planes of Fig. 5.11b, after carrying out one turn, *is actually double in width and has double the number of layers* (Fig. 5.12a).

We can expect that some scale invariance results from the infinite repetition of the process. At each turn the squeezing mechanism maps details of the attractor at a given scale to a smaller scale, again and again. Thus it should be possible to zoom on a very small part of the attractor and observe patterns that are also apparent at a larger scale. Such a scale invariance is typical of what is known as a *fractal*, a term coined by the mathematician Benoît Mandelbrot (1924–2010).

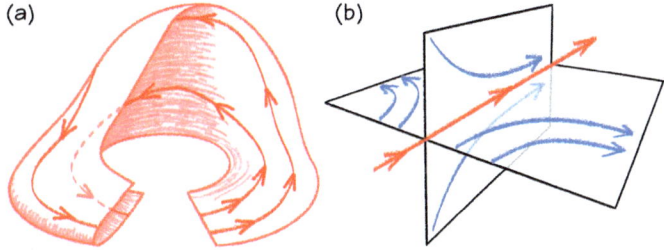

Figure 5.12: (a) Sketch of the geometrical transformation experienced by the section of the attractor as we wind around the vertical axis in Fig. 5.11a, summarizing the sequence of sections in Fig. 5.11b (based on (Abraham and Shaw, 1992)). (b) Structure of the velocity vector field around a typical trajectory, showing the unstable (here horizontal) and stable (here vertical) directions. In comparison to (a), the unstable direction is tangent to the sheet, and the stable direction is transverse to it.

5.3.3.b Stable and unstable directions

Let us examine more precisely how the stretching and folding mechanisms act along different directions. By closely observing the movie in Fig. 5.11b we can see that the Poincaré section is stretched longitudinally in the direction in which it already extends. The orbits located along the same segment of the attractor section diverge from each other: this is the unstable direction.

Then the section is folded over itself before being squeezed in a direction transverse to the stretching direction. Thus orbits whose separation is along this squeezing direction converge to each other: this is the stable direction.

Figure 5.12b depicts the corresponding structure of the flow around a typical trajectory, with some nearby trajectories converging along the stable direction, whereas most other nearby trajectories are influenced by the unstable direction and diverge from the reference trajectory. As we discussed in Sections 1.3.2.b and 2.3, around the trajectory, there exist a stable manifold and an unstable one that are tangent to the stable and unstable directions, respectively. The stable manifold of the trajectory consists of trajectories that converge to it forward in time, whereas the unstable manifold gathers trajectories that converge to it backward in time.

Our analysis suggests that the strange attractor is almost continuous along the unstable direction while consisting of an infinite stack of disjoint layers in the stable direction. This stacking is associated with the fractal structure.

5.3.4 Stretching and squeezing, a universal mechanism for generating chaos

Remarkably, the very same scenario is observed if the analysis is carried out on experimental signals, here measured in a CO_2 laser with modulated losses (Fig. 5.13). In fact, the stretching and folding mechanisms have been observed and characterized in countless different dynamical systems, both theoretical and experimental. This indicates that stretching, folding, and squeezing mechanisms are generic and universal for generating chaotic behavior.

To conclude, a phase space analysis of the Rössler system has allowed us to understand how it is possible to generate dynamical behavior that displays sensitivity to initial conditions in the sense that the distance between close trajectories diverges exponentially in time. This divergence results from the continuous stretching process, which precludes any forecast over a time interval larger than some characteristic time, which we call the *prediction horizon*. This expansion must be compensated by some contraction in the phase space so that phase space volumes remain finite, allowing the system to remain in a finite region of the phase space. The folding process, together with the squeezing that follows, ensures this by bringing back together trajectories that would otherwise flee to opposite ends of the universe.

Remarkably, nothing in the algebraic expression of the governing equations gives us a clue about this. It is the geometry of the velocity vector field that drives the orbits

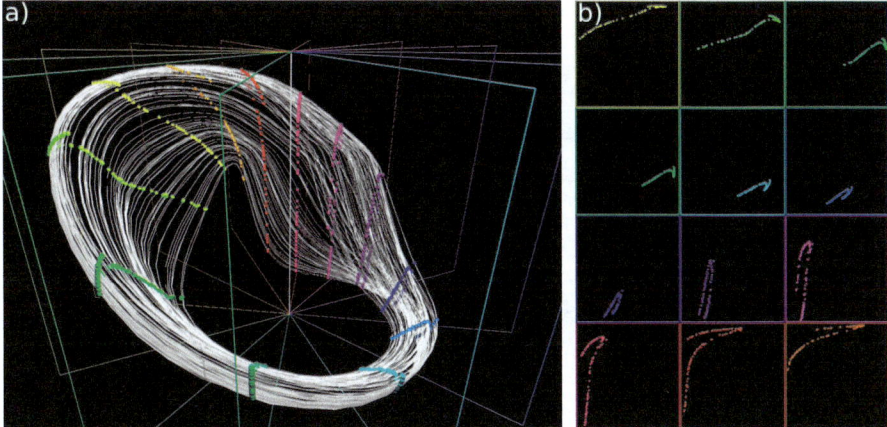

Figure 5.13: The data processing as in Fig. 5.11 is carried out for experimental signals coming from a CO_2 laser with modulated losses, showing the same stretching and folding processes operating to create the chaotic dynamics. (a) Strange attractor reconstructed in a phase space from the experimental signals; (b) Poincaré sections of the attractor obtained in successive section planes as one rotates around the vertical axis in (a).

so as to transform regions of the phase space as illustrated above for the Rössler system and the CO_2 laser.

This understanding of how geometrical mechanisms induce chaos will allow us to build quantitative measures of chaos. Estimating the stretching and squeezing rates will lead us to obtaining the so-called Lyapunov exponents. Not only their magnitude but also the number of positive exponents will inform us about the nature and intensity of chaos. Characterizing the infinitely foliated structure and the associated scale invariance will lead us to measure fractal dimensions, making us aware that strange attractors are complex geometrical objects of noninteger dimensions. These quantitative approaches to chaotic dynamics will be studied in the next chapter, Section 6.1.

5.3.5 The Smale horseshoe map and the essence of chaos

5.3.5.a The horseshoe map and its invariant set

In 1960 the American mathematician Stephen Smale imagined a simple geometrical dynamical system capturing the core mechanisms of the chaotic dynamics, stretching and squeezing (Smale, 1967). This system, known as the Smale horseshoe, provides us with a rationale for several paradoxical properties of chaos using a description in terms of symbolic dynamics.

Let us consider a unit square S and apply the horseshoe map f depicted in Fig. 5.14a, which consists of stretching, squeezing, and folding. Note that some points located inside the square escape it after one iteration. Thus the important question about the horseshoe

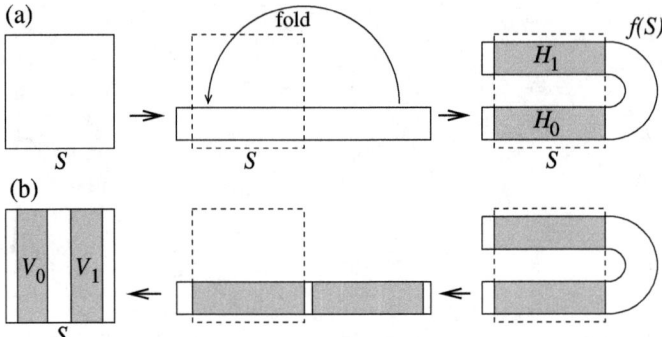

Figure 5.14: Geometrical description of the Smale horseshoe map: (a) The unit square is stretched hori-zontally and squeezed vertically, and then folded so as to intersect the original square along two horizontal bands H_0 and H_1; (b) The reverse transformation is applied on the horseshoe shape obtained, coming back to the original square. The preimages of the two horizontal bands $H_{0,1}$ are the vertical bands $V_{0,1}$.

map is the structure of its invariant set, which gathers points whose orbits remain in the square S forever:

$$S = \cap_{k=-\infty}^{\infty} f^k(S). \tag{5.4}$$

Because of its horseshoe shape, the image of the square intersects the original square along two horizontal bands so that we have (Fig. 5.14a)

$$S \cap f(S) = H_0 \cup H_1.$$

Now apply the reverse transformation to the image of the square so that we come back to the original square (Fig. 5.14b). We obtain two vertical bands V_0 and V_1, which are the preimages of the horizontal bands $H_{0,1}$, and thus $V_0 \cup V_1 = f^{-1}(S) \cap S$. Putting everything together, we see that

$$S \subset f^{-1}(S) \cap S \cap f(S) = \cup_{i,j} H_i \cap V_j.$$

Thus any point in the invariant set is located in one of the four regions $C_{i,j} \equiv H_i \cap V_j$ shown in Fig. 5.15, which we have labeled using a symbolic notation where the symbol before (resp., after) the dot tells us in which H_i (resp., V_i) we are. This forms the starting point of a powerful symbolic dynamical description of chaos.

We now follow the same step as before, considering the images $f(C_{i,j})$ and preim-ages $f^{-1}(C_{i,j})$ of the four components, keeping only what remains inside the square S (Fig. 5.16). We see that starting from the four squares of Fig. 5.16a, their four images are horizontal bands contained in a doubly folded region (Fig. 5.16b), whereas their four preimages are vertical bands contained in the original square (Fig. 5.16c).

In Fig. 5.16b, each horizontal band is tagged with the symbol sequence of the compo-nent it is coming from. However, we have shifted the two symbols completely left of the

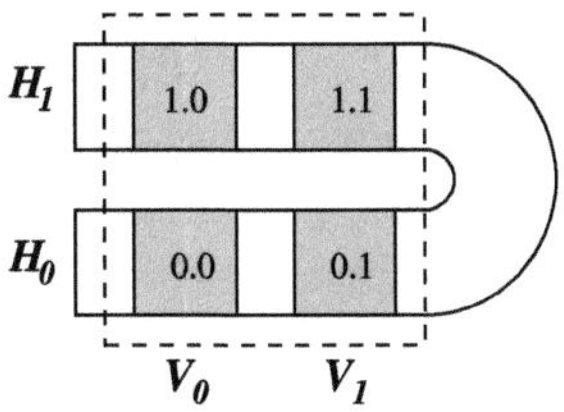

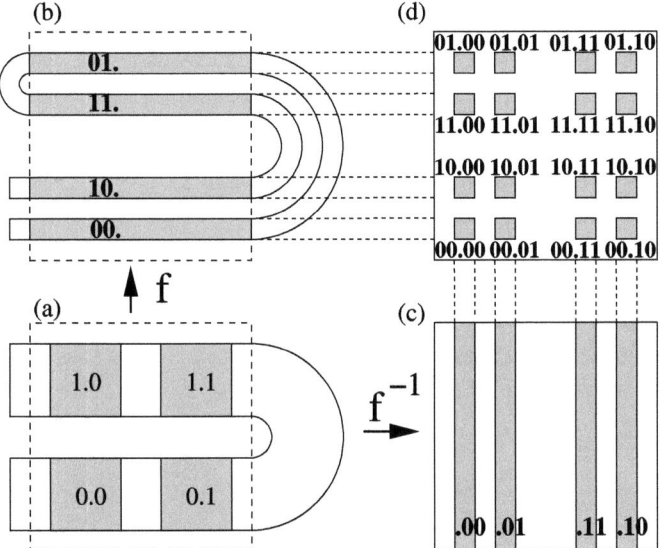

Figure 5.15: Since points in the invariant set S of the horseshoe transformation must belong to either H_0 and H_1, and simultaneously to either V_0 or V_1, they must fall into one of the four disjoint components shown here.

Figure 5.16: Starting from the four-component set in (a), we compute the intersections of its image with the original square in (b), and of its preimage with the square in (c). The 4 horizontal bands in (b) provide information about the vertical position, whereas the 4 vertical bands in (c) provide information about the horizontal position. By considering the intersections of the 4 horizontal and 4 vertical bands we obtain 16 components denoted by 4-digit sequences, two to the left of a central dot, and two to the right of it.

dot to indicate that these sequences only give us information about the vertical position, but not horizontally. Thus the horizontal bands are denoted $C_{ij\cdot}$.

Similarly, each vertical band in Fig. 5.16c is tagged with the symbol sequence of its preimage component, which however has been shifted completely to the right of the dot. This indicates that the sequence only provides us with information about the horizontal position of the band, but not vertically. The vertical bands are denoted $C_{\cdot ij}$.

Combining the two types of information, we can localize points in the invariant set both vertically and horizontally by considering the 16 components defined by

$$C_{ij\cdot kl} = C_{ij\cdot} \cap C_{\cdot kl},$$

as shown in Fig. 5.16d. The sequence to the right of the dot (called the forward sequence) localizes points along the horizontal direction, whereas the one to the left of the dot (called the backward sequence) localizes points along the vertical direction.

Recall that in this construction, the horizontal direction is the unstable direction, whereas the vertical one is the stable direction. Hence the forward (resp., backward) sequence informs us about the position along the unstable (resp., stable) direction.

Now we can imagine repeating this construction ad libitum and obtaining an infinite number of disjoint components with zero diameter (i. e., points), labeled by two infinite sequences separated by a dot. We see that the invariant set of the horseshoe map has a very complicated structure, which we will characterize now using a symbolic dynamical approach that capitalizes on our analysis.

This leads us to a correspondence between infinite two-sided sequences of 0 and 1 and points in the invariant set, which we will explore further below.

5.3.5.b Unfolding the complexity of chaos with symbolic dynamics
5.3.5.b-i From points to symbol sequences

Looking at Figs. 5.16d, we see that the symbol immediately to the right of the dot informs us on whether we are in V_0 or in V_1. We also see that when we compute the image of a component, this symbol goes to the left (compare Figs. 5.16c and 5.16a), whereas it goes to the right when we compute the preimage (compare Figs. 5.16b and 5.16a).

This inspires us to consider the following symbolic coding:

$$s(x) = \begin{cases} 0 & \text{if } x \in V_0, \\ 1 & \text{if } x \in V_1 \end{cases}$$

and to associate the following biinfinite sequence with any point in the invariant set of the horseshoe map:

$$\Sigma(x) = \ldots s_{-j}s_{-j+1} \ldots s_{-2}s_{-1} \cdot s_0 s_1 s_2 \ldots s_i s_{i+1} \ldots \quad \text{with } s_i = s(f^i(x)), \quad (5.5)$$

where we keep track of whether iterates and preiterates of x fall inside V_0 or V_1. Actually, this is the generalization of the coding illustrated in Fig. 5.16.

Looking at Fig. 5.16d, we can convince ourselves that if we have two points whose sequences are identical for the first i symbols along the forward sequence and for the first j symbols along the backward sequence (but not more):

$$\Sigma(x) = \ldots s_{-j-1}s_{-j} \ldots s_{-2}s_{-1} \cdot s_0 s_1 s_2 \ldots s_{i-1}s_i \ldots \quad \text{and}$$
$$\Sigma(x') = \ldots s'_{-j-1}s_{-j} \ldots s_{-2}s_{-1} \cdot s_0 s_1 s_2 \ldots s_{i-1}s'_i \ldots , \quad (5.6)$$

the two points are very close to each other, as they belong to the same component $C_{s_{-j}\ldots s_{-2}s_{-1} \cdot s_0 s_1 s_2 \ldots s_{i-1}}$. We can show that there exist α, β such that the distance between the two points satisfies

$$\frac{\beta}{2^k} < d(x, x') < \frac{\alpha}{2^k}, \quad \text{where } k = \min(i, j). \tag{5.7}$$

We thus have the remarkable property that if two points have the same sequence, then their distance is zero, and therefore they are the same point. Since any sequence corresponds to one and only one point, namely $\{x\} = \cap_{k \geq 1} C_{s_{-k} \ldots s_{-2} s_{-1} \cdot s_0 \ldots s_{k-1}}$, and since any point in the invariant set can be associated with a sequence, there is a bijection between points of the invariant set and biinfinite sequences of 0 and 1.

In fact, the knowledge of the infinite sequence associated with a point allows us to localize it with arbitrary precision in the plane. The forward sequence $\Sigma_+ = s_0 s_1 s_2 \ldots$ indicates the position along the unstable direction, whereas the backward sequence $\Sigma_- = s_{-1} s_{-2} s_{-3} \ldots$ indicates the position along the stable direction.

Property (5.7) is very important because it will allow us to find points that approach a given point arbitrarily close by finding points whose sequences agree with that of the given point over sufficiently many symbols around the dot.

Definition (5.5) implies that

$$\Sigma(f(x)) = \ldots s_{-j} s_{-j+1} \ldots s_{-2} s_{-1} s_0 \cdot s_1 s_2 \ldots s_i s_{i+1} \ldots = \sigma \Sigma(x),$$

where the shift operator σ shifts the sequence one symbol to the left.

To summarize, we have a one-to-one correspondence between points in the invariant sets and symbol sequences, and applying the horseshoe transformation in space is equivalent to shifting the symbol sequence

$$\Sigma(f(x)) = \sigma \Sigma(x) \tag{5.8}$$

so that the horseshoe dynamics is faithfully represented in the space of sequences.

We can now use this fact to unveil some remarkable properties of chaotic behavior.

5.3.5.b-ii The fundamental properties of chaos

Using symbolic dynamics, we can prove the following properties of chaotic behavior.

- **Sensitivity to initial conditions.** Take arbitrarily close points x and x' such as in (5.6) by taking i and j as large as needed. By shifting the two sequences sufficiently many times we can obtain that their leading symbols differ: $s(x) \neq s(x')$. Their distance is macroscopic since they are in different V_i bands. Moreover, (5.7) implies that their distance is multiplied, on average, by two at each step.
- **Existence of an infinity of periodic orbits of arbitrarily high period**. Take any infinite sequence that is the repetition of a finite symbol string

$$\Sigma(x) = (s_0 s_1 s_2 \ldots s_{p-1})^\infty \cdot (s_0 s_1 s_2 \ldots s_{p-1})^\infty,$$

where p is the period. Then $\sigma^p \Sigma(x) = \Sigma(x)$, and because of the bijection between points and sequences, $f^p(x) = x$. Since there are infinitely many periodic sequences and since they can have arbitrarily high period, the same is true for periodic points in the original space.

– **Periodic orbits are dense in the invariant set**. Given a point X in the invariant set, we consider the finite sequence $s_{-j} \ldots s_{-2} s_{-1} \cdot s_0 s_1 s_2 \ldots s_{i-1}$ made by the first j symbols of its backward sequence and the first i symbols of its forward sequence. By extending this sequence periodically backward and forward we obtain a periodic orbit that approaches the point as specified by (5.7). By choosing arbitrarily large i and j we can find periodic orbits arbitrarily close to the point X. Hence periodic orbits are dense.

– **Existence of a dense orbit** (an orbit that comes arbitrarily close to any point in the invariant set). Construct a special infinite sequence that contains all possible finite sequences of 0 and 1. By iterating the shift operator on it we can arrange to obtain any possible combination for the first j symbols of the backward sequences and for the first i symbols of the forward sequence, thus approaching any point within a distance bounded by inequalities (5.7). Since i and j can be made as large as needed to make this distance as small as wanted, this shows that any point in the invariant set will be approached arbitrarily closely by the special orbit. The existence of a dense orbit guarantees that the invariant set cannot be decomposed into smaller invariant sets.

These properties are typical of a chaotic dynamics and can even be used to give a rigorous definition of it.

5.3.6 The Hénon map

The horseshoe map is well suited to elaborate mathematical proofs, but it is not a realistic dynamical system as such because most trajectories leave the square. Actually, it has been shown that horseshoe maps can be identified in Poincaré maps of systems where the unstable and stable manifolds have developed transverse intersections, but we will not touch this complex subject here.

To study a realistic once-folding map without having to resort to the Poincaré map of a flow, the French astronomer Michel Hénon (1931–2013) designed the following two-dimensional recurrence system (Hénon, 1976), here given in a slightly different form from the original:

$$\begin{cases} x_{n+1} = 1 + by_n - ax_n^2, \\ y_{n+1} = x_n, \end{cases} \tag{5.9}$$

where a controls the nonlinearity, and b controls the dissipation.

Figure 5.17a shows the square $S = [-2, 2] \times [-2, 2]$ with its images $f(S)$ and $f^2(S)$ under the Hénon map f (Eq. (5.9)) and should be compared with Fig. 5.16. The folding mechanism is clearly discernible although we can see on $f^2(S)$ that it is not complete: at some places, there are four layers, and at some others, there are only two. This indicates that the symbolic dynamics is not complete (not all sequences of 0s and 1s are present), which is generally the case for real attractors. A picture of the Hénon attractor is obtained by iterating the recurrence (5.9) over a long orbit (Fig. 5.17b).

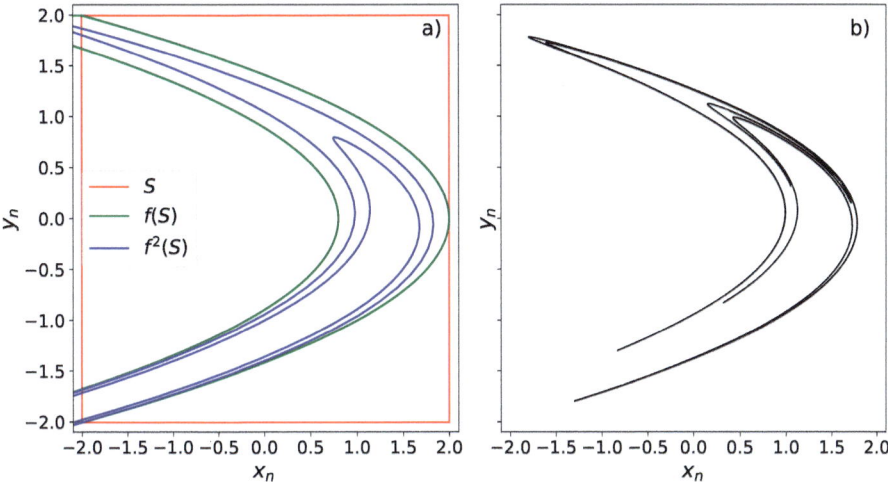

Figure 5.17: (a) The square $S = [-2, 2] \times [-2, 2]$ is shown together with its iterate and its second iterate under the Hénon map (5.9) (b) Attractor of the Hénon map (5.9) computed for $a = 1.4$ and $b = 0.3$.

Note that as $b \to 0$, the recurrence system (5.9) reduces to a one-dimensional map. At $b = 0.3$ the dissipation rate is moderate, allowing us to observe a complex structure, which we will later characterize as a fractal.

5.4 Dealing with real data

5.4.1 Reconstruction of an attractor from experimental signals

When studying the Rössler system, we have shown that much information about a chaotic regime can be obtained from the corresponding attractor in the phase space. We did so using the complete set of natural variables, namely x, y, and z.

When we observe an experimental system, however, we generally have access to a very limited number of variables, most often a single one. Taking the example of a laser, the output intensity is the only easily measurable quantity. Moreover, there is actually no clear definition of what are good variables for the system, since this depends in great part on how we model it. Many relevant variables may forever remain unknown to us. Thus we should focus on measuring quantities that do not depend on the coordinates chosen, which implies that they are invariant under a change of coordinates.

The question then arises as to how characterize a chaotic regime when only one irregular time series $X(t)$ is known, provided of course that we know that this variable is representative of the dynamics of the system. If we are to carry out a phase space approach, then the question is what would make a good state for the system?

This makes us return to the definition of a dynamical state as a collection of data from which the future time evolution can be predicted, as we discussed in Section 1.1.2. Returning to the concepts we elaborated then and examining again formula (1.3), we know that $X(t), \dot{X}(t), \ddot{X}(t), \ldots$ are natural state variables for predicting the time evolution of $X(t)$.

In the case of the Rössler system, it is even possible to show that the change of coordinates from (x, y, z) to $(y, \dot{y}, \ddot{y})$ is a bijection, and thus preserves information about the state. Computing the successive derivatives of y, we find that

$$\begin{cases} y = y, \\ \dot{y} = x + ay, \\ \ddot{y} = \dot{x} + a\dot{y} = ax + (a^2 - 1)y - z, \end{cases}$$

and thus we have

$$\begin{pmatrix} y \\ \dot{y} \\ \ddot{y} \end{pmatrix} = \begin{pmatrix} 0 & 1 & 0 \\ 1 & a & 0 \\ a & a^2 - 1 & -1 \end{pmatrix} \begin{pmatrix} x \\ y \\ z \end{pmatrix},$$

where the matrix involved in the coordinate change has determinant 1 and thus is never singular. Note that if we try to use z as a unique variable, then the change of coordinate is singular in the plane $z = 0$ and thus is not suitable.

However, it is generally not practical to use time derivatives when analyzing experimental data, because numerical differentiation is very sensitive to noise. Indeed, it overemphasizes the high-frequency components of the signal, which are generally strongly contaminated by noise. An alternative and very popular method is to use time-delayed coordinates $\{X(t), X(t + \tau), X(t + 2\tau), \ldots, X(t + (n-1)\tau)\}$, where τ is a fixed delay. Considering the Taylor expansions of these time-delayed coordinates,

$$X(t) = X(t),$$

$$X(t + \tau) = X(t) + \tau \dot{X}(t) + \frac{\tau^2}{2} \ddot{X}(t) + \cdots,$$

$$X(t + 2\tau) = X(t) + 2\tau \dot{X}(t) + 2\tau^2 \ddot{X}(t) + \cdots,$$

we see that they provide us with a change of coordinates between the time-derivative and time-delayed coordinates. Reconstructed attractors using one of those coordinate systems are called *embeddings* since the original attractor (assuming that it can be defined as such) is embedded in the reconstruction space.

In the case of modulated systems described by nonautonomous equations, we easily obtain an additional state variable, which is the phase of modulation. An attractor reconstructed from signals from a modulated CO_2 laser is shown in Fig. 5.18.

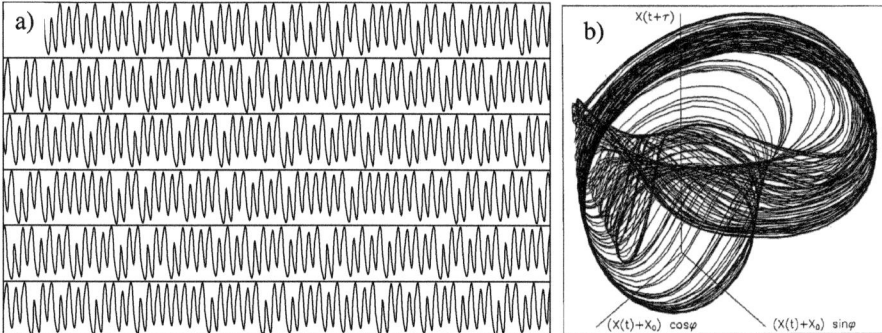

Figure 5.18: (a) Chaotic time series $X(t)$ delivered by a CO_2 laser with modulated losses; (b) reconstruction of the underlying strange attractor in a phase space with cylindrical coordinates $\{X(t), X(t + \tau), \phi\}$, where τ is a suitably chosen time delay, and ϕ is the modulation phase. Reprinted from (Lefranc and Glorieux, 1993).

In practice, we have to choose the number of coordinates we use, which is called the *embedding dimension* (or dimension of the embedding space). It must be chosen so that trajectories do not cross in the phase space, which is a necessary condition for it to be considered a space of states, where the future is determined from the initial condition. To simplify visualization, it makes sense to choose the smallest dimension that satisfies this constrains, but for some applications, using higher values of the embedding dimension may be beneficial. Looking closely at the reconstructed attractor from Fig. 5.18, we can indeed see that neighbor points have almost identical velocity vectors (Fig. 5.19), which is an indication of the deterministic nature of the dynamics since we can assume a relation of type (1.5).

For time-delayed embeddings, we have additionally to fix the value of the time delay τ, which should be such that the different time-delayed values bring independent information. Thus τ should not be too small, or the variables will be too close. It should not be too large, or the variables will be decorrelated, and we get noise. In general, a good order of magnitude is a fraction of a characteristic time of the system. For example, if T is the average period of rotation around the vertical axis in Fig. 5.11, then a good starting value would be $T/4$ or $T/5$. Then we proceed by trial and error to find a value around which the quantity measured does not vary much. The choice is not very critical when the value is in the good range.

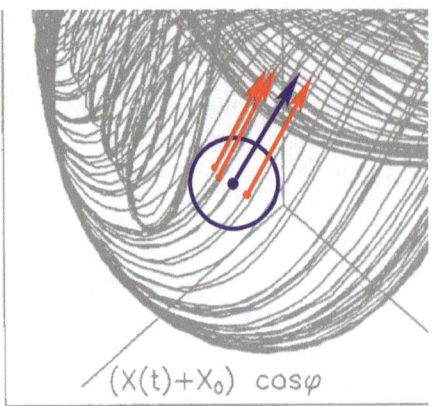

$(X(t)+X_0) \ \cos\varphi$

Figure 5.19: When a reconstruction of a chaotic attractor (here the one of Fig. 5.18) is a good embedding, neighbor points in the phase space have almost identical velocity vectors.

5.4.2 Computing a return map from the maxima of a time series

When a strange attractor has been embedded in a reconstructed phase space, one of the first steps to analyze it is generally fixing a Poincaré section and computing the associated first return map. This allows us to verify whether the dynamics is deterministic, just as we discussed for the Rössler system in Fig. 5.10. As with the latter, this method is especially useful when the system is very dissipative and the Poincaré section is close to being one-dimensional, making it possible to represent the return map as a map of an interval into itself.

When we analyze the time series of a single variable $x(t)$, as it is often the case experimentally, there is a Poincaré section that is particularly easy to perform: the one corresponding to $\dot{x} = 0$ and $\ddot{x} < 0$, which corresponds to picking all the maxima x_i^{max} of the time series. This is an implicit time-derivative embedding, but since we do not compute explicitly the derivative, this is not a problem. Computing the return map then consists in plotting the $(i + 1)$th local maximum x_{i+1}^{max} vs. the previous one x_i^{max}.

Using this simple method, we can, for example, obtain the first-return map for the attractor observed by Hudson and Mankin (1981) in a chemical reaction (Fig. 5.20). The fact that the graph is that of a single-valued function shows clearly the deterministic nature of the dynamics. As with the Rössler attractor, we observe that the return map is nonbijective, again pointing at the action of stretching and squeezing.

When we do not obtain a simple one-dimensional plot, because the system is not dissipative enough, it is always possible to plot a Poincaré section with coordinates $(x(t_i), x(t_i + \tau))$, where t_i are the times of the local maxima, and τ is suitably chosen.

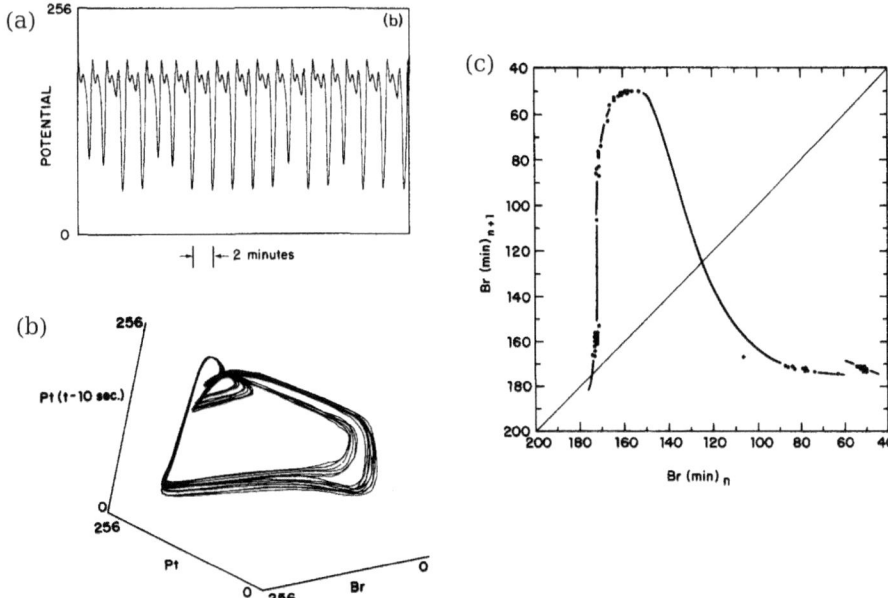

Figure 5.20: Reprinted from (Hudson and Mankin, 1981) with the permission of AIP Publishing. (a) Voltage measured from bromide ion electrode showing aperiodic dynamics of the concentration of this ion. (b) Reconstructed attractor from the measurements of two electrodes (platinum wire electrode *Pt* and bromide ion electrode *Br*). For the *z*-axis time delay coordinate $Pt(t - 10s)$ is used. (c) First return map obtained by plotting each local minima of the *Br* time series as a function of the previous one (the solid line is a fit).

5.5 Conclusions

In this chapter, we studied two dynamical regimes that are observed in dimension three and above. Quasi-periodicity results from the combination of two or more oscillations and is confined to a regular surface in the phase space, an invariant torus. Depending on the ratio of the interacting frequencies, the trajectory fills the entire torus or follows a periodic orbit embedded in it. In both cases, quite complex waveforms can be generated, especially if many frequencies are involved.

Deterministic chaos also generates irregular signals; however, it has a clear signature, sensitivity to initial conditions. This requires the dynamics to unfold in more than two dimensions to circumvent the no-crossing theorem. The geometrical mechanisms generating chaos in the phase space are stretching, folding, and squeezing, and their combined action results in attractors of fractal structure. The way in which these mechanisms operate leads us to a symbolic dynamical analysis of chaos that reveals its essential properties.

Using embedding techniques, attractors can be reconstructed from experimental signals and can thus be characterized, as we will see in Chapter 6.

Exercises

Rössler model

Consider the Rössler system

$$\begin{cases} \dot{x} = -y - z, \\ \dot{y} = x + ay, \\ \dot{z} = b + z(x - c), \end{cases}$$

with $a = b = 0.2$ and $c = 5.7$.

1. Integrate numerically the system and draw the strange attractor.
2. Perform a Poincaré section using the plane $x = 0$ with $\dot{x} > 0$.
3. Draw the first return map using the method of Section 5.3.2.
4. Explore different regimes of the Rössler model by varying the parameters a, b, and c.

Lorenz attractor

Consider the Lorenz system

$$\begin{cases} \dot{X} = \Pr(Y - X), \\ \dot{Y} = -XZ + rX - Y, \\ \dot{Z} = XY - bZ, \end{cases}$$

with $\Pr = 10$, $b = 8/3$, and $r = 28$.

1. Integrate numerically the system and observe the aperiodic behavior of the solutions. Represent an asymptotic trajectory in the phase space.
2. Perform a Poincaré section of the attractor using the plane $Z = r - 1$ with $\dot{Z} > 0$.
3. Draw the first return map using the method of Section 5.4.2, i. e., by plotting the local maximum Z_{i+1} as a function of the previous one Z_i. Discuss the stretching and folding mechanisms.

Forced van der Pol oscillator in the phase-locked regime

Consider the driven van der Pol oscillator of Eq. (5.1)

$$\begin{cases} \dot{X}_1 = X_2, \\ \dot{X}_2 = -\varepsilon(X_1^2 - 1)X_2 - X_1 + \Gamma \cos X_3, \\ \dot{X}_3 = \omega. \end{cases}$$

1. Integrate numerically the system for $\varepsilon = 33$, $\omega = 1$, and $\Gamma = 3.3$. Draw an asymptotic trajectory in the phase space.
2. Perform a Poincaré map using the stroboscopic method and show that the system displays a 1:9 frequency locking.
3. The same questions for $\varepsilon = 10$, $\omega = 1$, and $\Gamma = 4.75$, corresponding to a 1:3 frequency locking.

Other locking parameters can be found in (Flaherty and Hoppensteadt, 1978).

Symbolic dynamics and periodic orbits

In Section 5.3.5, we saw that a description of chaos in terms of a symbolic dynamics is possible. In the case of a horseshoe-like once-folding system, an orbit is associated with a sequence of 0s and 1s. Thus periodic orbits of the system are associated with a periodic symbol sequence, which allows us to enumerate all possible periodic orbits.

Assuming that all sequences of 0s and 1s are possible:
1. For periods up to 10, find the number of periodic points of each period (which is equal to the number of sequences of 0s and 1s).
2. Count those points whose sequence is the repetition of a shorter sequence and has actually a lower period.
3. Given that a period-p orbit has p periodic points, count the number of genuine period-p orbits.

Symbolic dynamics and topological entropy

Again assume a symbolic dynamics where each finite sequence of 0s and 1s can be observed.
1. Count the number $\mathcal{N}(l)$ of different symbolic sequences of length l.
2. Defining the topological entropy as $h_T = \lim_{l\to\infty} \frac{\ln \mathcal{N}(l)}{l}$, compute the topological entropy of the horseshoe.

6 Deeper into chaos

In the previous chapter, we described qualitatively the structure of strange attractors to understand the mechanisms explaining how irregular and unpredictable time series can emerge from low-dimensional deterministic systems. However, chaos theory is more than qualitative. It also provides us with tools to quantify chaotic dynamics, and more importantly, it also reveals how some features of chaotic dynamics are universal, allowing us to derive general conclusions from the study of specific models.

In this chapter, we address two fundamental aspects of chaos theory: (1) how to compute quantitative invariants of chaos; (2) the structure of one of the universal roads to chaos, the period-doubling cascade.

6.1 Quantitative characterization of strange attractors

In Section 5.3.3, we showed that chaotic dynamical regimes result from the interplay of stretching and folding mechanisms in the phase space and that they explain the irregular behavior and sensitivity to initial conditions that are observed. Here we introduce two measures of chaos that quantify these two complementary mechanisms:

- The Lyapunov exponents characterize the time evolution of trajectories in a close neighborhood of a reference trajectory, measuring the stretching and squeezing rates along different directions. In particular, the distance d between two trajectories with almost identical initial conditions typically evolves as $d(t) \sim d(0)e^{\lambda_1 t}$, where λ_1 is the largest Lyapunov exponent.
- Various extensions of the usual concept of dimension indicate us to which extent a strange attractor is fractal and, in particular, how the different layers stacked by the folding process are organized.

The two measures are not totally independent, as the Lyapunov exponents can be used to compute the Lyapunov dimension, which is an estimate of fractal dimension.

6.1.1 Lyapunov exponents

6.1.1.a Introduction
A main feature of chaotic dynamics on a strange attractor is its sensitivity to initial conditions, which stems from the stretching mechanism continuously acting on neighboring trajectories, and which makes the system asymptotically unpredictable. In practice, it is important to know the characteristic time beyond which no prediction is possible. More generally, we have seen that stretching in some directions is compensated by squeezing in some others, and we may be interested to put numbers on these different geometrical mechanisms.

https://doi.org/10.1515/9783110677874-006

These questions can be answered using the concept of Lyapunov exponents. Before describing the theory, let us look more closely at the phenomenon we want to describe. In Fig. 6.1, we follow the time evolution of a bunch of 1000 different states of the Rössler system, all of them close to each other by about one percent in phase space. At first, the representative points remain close, rotating at about the same speed around the origin, but then they separate more and more rapidly along a one-dimensional curve (Figs. 6.1a–d). Then they lose their phase coherence and begin to spread all over the attractor. In Fig. 6.1f, we see that the system can be anywhere on the attractor. Thus the system is unpredictable over that time scale. During the entire movie, the bunch of states remains stable in the direction transverse to the attractor due to squeezing. Moreover, the relative maintaining of phase coherence at the beginning of the time evolution indicates that there is a neutral direction in the direction of the flow.

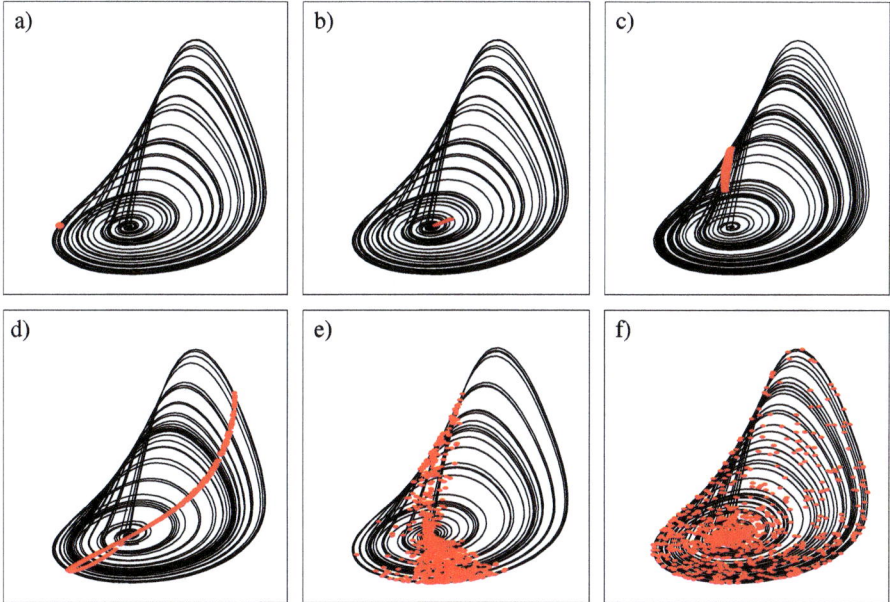

Figure 6.1: Time evolution of a bunch of states of the Rössler system for $(a, b, c) = (0.434, 2.0, 4.0)$. Starting from $t = 0.0$ in a), the configuration of the bunch is shown at b) $t = 10.0$; c) $t = 20.0$; d) $t = 30.0$; e) $t = 100.0$; f) $t = 1000.0$.

As a first step, characterizing the exponential divergence of trajectories can be done in a relatively simple way. Consider a trajectory $\mathbf{X}_1(t)$ on the attractor, originating from the initial condition $\mathbf{X}_1(0)$. Take a second trajectory $\mathbf{X}_2(t)$ starting at $\mathbf{X}_2(0)$ very close to $\mathbf{X}_1(0)$. Now monitor $\delta\mathbf{X}(t) = \|\mathbf{X}_2(t) - \mathbf{X}_1(t)\|$, the distance between the two trajectories at time t (Fig. 6.2).

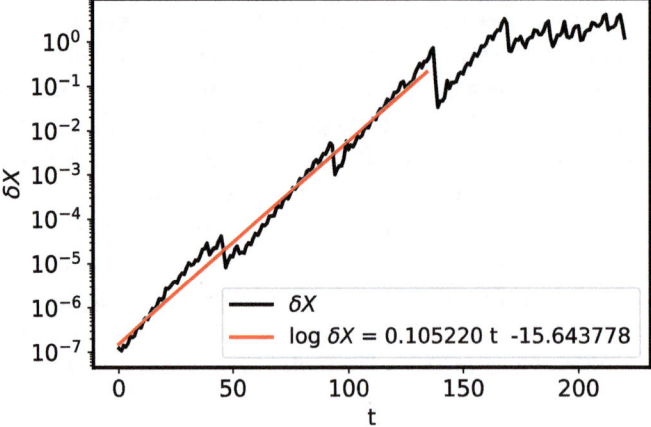

Figure 6.2: Evolution with time of the distance between two neighboring trajectories of the Rössler attractor. This distance first grows exponentially fast (note the logarithmic scale along the vertical axis) and then saturates when the distance becomes comparable to the diameter of the attractor.

When small enough, this distance behaves like $\delta\mathbf{X}(t) = \delta\mathbf{X}(0)e^{\lambda_1 t}$, where $\lambda_1 \simeq 0.105\ldots$ is the largest Lyapunov exponent. After some time, the distance is saturated by the diameter d_{max} of the attractor, indicating that the two trajectories can be anywhere on the attractor. If ϵ is the typical resolution with which we can resolve a state (due to noise or technical uncertainties), then the prediction horizon can be defined as the time over which an error of ϵ becomes macroscopic:

$$T_{predict} = \frac{1}{\lambda_1} \ln \frac{d_{max}}{\epsilon}.$$

Interestingly, increasing the resolution by a factor of two ($\epsilon_{new} = \epsilon/2$) only increases the horizon time by a modest amount $\Delta T_{predict} = \ln 2/\lambda_1 \simeq 6.6$ time units. This explains why it is difficult to predict the behavior of chaotic systems over long times (and, incidentally, why weather prediction is so hard, the dynamics of the atmosphere being indeed chaotic).

6.1.1.b Mathematical definition of the Lyapunov exponents

We noted previously that the dynamical behavior around a reference trajectory varies in different directions of the phase space, as it can display stretching, neutrality, or squeezing (see Fig. 5.12b). We thus need to characterize the relative evolution of two trajectories not only in a single direction, but in the whole neighborhood.

In Section 2.4, we already studied the problem of determining the time evolution of a perturbation $\delta\mathbf{X}(t)$ around a reference trajectory $\mathbf{X}_{ref}(t)$ (see Fig. 2.9). We found that the perturbation satisfies the following equation (reproducing Eq. 2.12):

$$\frac{\mathrm{d}\delta\mathbf{X}(t)}{\mathrm{d}t} = \left(\frac{\partial\mathbf{F}}{\partial\mathbf{X}}\right)\bigg|_{\mathbf{X}=\mathbf{X}_{\mathrm{ref}}(t)} \delta\mathbf{X}(t),$$

where $(\partial\mathbf{F}/\partial\mathbf{X})|_{\mathbf{X}=\mathbf{X}_{\mathrm{ref}}(t)}$ is the Jacobian of the flow. We noted that if we know the fundamental matrix solution of the above equation (reproducing Eq. (2.13))

$$\frac{\mathrm{d}}{\mathrm{d}t}\mathbf{M}(t) = \left(\frac{\partial\mathbf{F}}{\partial\mathbf{X}}\right)\bigg|_{\mathbf{X}=\mathbf{X}_{\mathrm{ref}}(t)} \mathbf{M}(t), \quad \mathbf{M}(t_0) = \mathbb{1},$$

then the solution of the variational equation is

$$\delta\mathbf{X}(t_0 + \tau) = \mathbf{M}(\tau)\delta\mathbf{X}(t_0),$$

and thus the long-term behavior of a perturbation is governed by the properties of the matrix $\mathbf{M}(t)$. If we are only interested in how the magnitude of the perturbation evolves, then it is advantageous to consider

$$\left\|\delta\mathbf{X}(t_0 + \tau)\right\|^2 = \delta\mathbf{X}(t_0)\mathbf{M}(\tau)^T.\mathbf{M}(\tau)\delta\mathbf{X}(t_0) = \left\|\mathbf{M}(\tau)\delta\mathbf{X}(t_0)\right\|^2$$

because the matrix $\mathbf{M}(t)^T\mathbf{M}(t)$ is symmetric and thus can always be diagonalized and has orthogonal eigenvectors. It can be shown that by diagonalizing this matrix we obtain

$$\mathbf{M}(t)^T\mathbf{M}(t) \sim \mathbf{P}\begin{pmatrix} e^{2\lambda_1 t} & & & 0 \\ & e^{2\lambda_2 t} & & \\ & & e^{2\lambda_i t} & \\ 0 & & & e^{2\lambda_n t} \end{pmatrix}\mathbf{P}^{-1}, \tag{6.1}$$

where $\mathbf{P}$ is the matrix describing the change of basis needed for diagonalization, and $\lambda_1 > \lambda_2 > \lambda_3 > \cdots > \lambda_n$ are the Lyapunov exponents. It should not be a surprise that the diagonal matrix elements consist of exponential functions if we remember that this is already the case where the Jacobian matrix is constant (Section 2.2.1.b).

The Lyapunov exponents can therefore be obtained from

$$\lim_{t\to\infty}\frac{1}{2t}\ln\mathbf{M}(t)^T\mathbf{M}(t) = \mathbf{P}\begin{pmatrix} \lambda_1 & & & 0 \\ & \lambda_2 & & \\ & & \lambda_i & \\ 0 & & & \lambda_n \end{pmatrix}\mathbf{P}^{-1}, \tag{6.2}$$

where the logarithm of a matrix is defined by its series expansion, and each exponent is associated with a different direction of the orthogonal basis. Formula (6.2) expresses that the Lyapunov exponents are the eigenvalues of the left-hand term, the eigenvectors of which are given by the columns of the matrix $\mathbf{P}$.

Figure 6.3 summarizes the principle of computation. As we integrate the variational equation and solve Eq. (6.2), we obtain eigenvalues and eigenvectors of $\mathbf{M}(t)^T\mathbf{M}(t)$. The

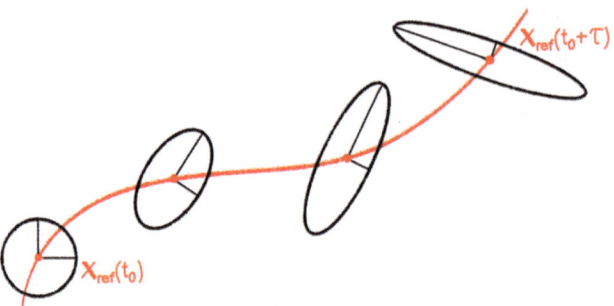

Figure 6.3: Computation of the Lyapunov exponents as defined by Eqs. (6.1) and (6.2). The ellipses show the directions in which the norm of an initially isotropic perturbation has increased or decreased.

eigenvectors are orthogonal because $\mathbf{M}(t)^T\mathbf{M}(t)$ is symmetric. The associated eigenvalues indicate the amount of stretching/squeezing along the given direction. Typically, the sum of the k largest Lyapunov exponents give the growth rate of k-dimensional volume elements, which align along the subspace spanned by the eigenvectors associated with these Lyapunov exponents.

6.1.1.c Practical determination of Lyapunov exponents

In practice, the computation of Lyapunov exponents is faced with two related difficulties: the matrix $\mathbf{M}(t)$ tends to become singular because it is dominated by its largest eigenvalue (all perturbations align along the most unstable direction), and its norm grows without bound.

Usually, the computation is carried out in finite time steps, which can correspond to integrating the equations over a fixed time interval or to the time of flight between successive returns in a Poincaré section. Thus we have a series of points $\mathbf{X}_i$ that are related by $\mathbf{X}_i = \phi(\mathbf{X}_{i-1})$, where ϕ can be considered as fixed-time flow map or as a Poincaré return map. Given a perturbation $\delta\mathbf{X}_i$ around $\mathbf{X}_i$, we write $\mathbf{L}_{\mathbf{X}_i} = (\partial\phi/\partial\mathbf{X})_{\mathbf{X}=\mathbf{X}_i}$, the linearized flow map around $\mathbf{X}_i$ such that $\delta\mathbf{X}_{i+1} = \mathbf{L}_{\mathbf{X}_i}\delta\mathbf{X}_i$.

When we analyze experimental data, we do not have the governing equations. Hence we have to estimate the matrices $\mathbf{L}_{\mathbf{X}_i}$ numerically. This is usually done by collecting a set of neighboring points around $\mathbf{X}_i$, monitoring how their trajectories evolve until next step, and adjusting a numerical matrix $\mathbf{L}_{\mathbf{X}_i}$ to these data. However, it is generally difficult to estimate reliably the part corresponding to the most negative exponents, because the extension of the attractor is very small in these directions and is thus contaminated by noise.

Starting from an initial condition $\mathbf{X}_0$, a naive calculation of $\mathbf{M}(t)$ would be to carry out the following steps:

$$\mathbf{M}_0 = \mathbb{1},$$
$$\mathbf{M}_1 = \mathbf{L}_{\mathbf{X}_0}\mathbf{M}_0,$$

$$\mathbf{M}_2 = \mathbf{L}_{\mathbf{X}_1}\mathbf{M}_1,$$

$$\cdots$$

$$\mathbf{M}_p = \mathbf{L}_{\mathbf{X}_{p-1}}\mathbf{M}_{p-1}.$$

However, the matrices $\mathbf{M}$ would suffer from the above-mentioned problem: all their columns become aligned with the most unstable direction, making it impossible to recover anything except the largest Lyapunov exponent. A clever algorithm was designed to overcome this problem (Eckmann et al., 1986), which consists in maintaining an orthogonal frame on which we apply the linearized maps.

The algorithm makes use of the QR decomposition, where a matrix $\mathbf{A}$ can be written as $\mathbf{A} = \mathbf{QR}$ with orthogonal matrix $\mathbf{Q}$ and upper triangular matrix $\mathbf{R}$, for which powerful numerical algorithms exist. If the columns $\mathbf{v}_i$ of $\mathbf{A}$ are viewed as a basis of vectors, then the columns $\mathbf{w}_i$ of $\mathbf{Q}$ are the orthonormal basis obtained by Gram–Schmidt orthogonalization. The first vector $\mathbf{w}_1 = \mathbf{v}_1/|\mathbf{v}_1|$ is $\mathbf{v}_1$ normalized to unity, the second vector $\mathbf{w}_2 = (\mathbf{v}_2 - \mathbf{v}_2 \cdot \mathbf{w}_1)/|\mathbf{v}_2|$ is the normalized second basis vector with any component along $\mathbf{w}_1$ removed. The vector $\mathbf{w}_i$ is the normalized projection of $\mathbf{v}_i$ onto the space orthogonal to $\mathbf{w}_1,\ldots,\mathbf{w}_{i-1}$, etc. In this way, the first basis vector will always be aligned to the most unstable direction, the second basis vector will be the remaining most unstable direction in the orthogonal space to $\mathbf{w}_1$, and so on.

The sequence now reads (where $\mathbf{A} = \mathbf{QR}$ means: given $\mathbf{A}$, compute $\mathbf{Q}$ and $\mathbf{R}$ with the QR algorithm and make the result $\mathbf{Q}$ available to the next step of algorithm)

$$\mathbf{Q}_0 = \mathbb{1},$$
$$\mathbf{L}_{\mathbf{X}_0}\mathbf{Q}_0 = \mathbf{Q}_1\mathbf{R}_1,$$
$$\mathbf{L}_{\mathbf{X}_1}\mathbf{Q}_1 = \mathbf{Q}_2\mathbf{R}_2,$$
$$\cdots$$
$$\mathbf{L}_{\mathbf{X}_{p-1}}\mathbf{Q}_{p-1} = \mathbf{Q}_p\mathbf{R}_p.$$

At the end of the computation, the columns of $\mathbf{Q}_p$ are the normalized eigenvectors of $\mathbf{M}^T\mathbf{M}$, whereas the diagonal elements of the $\mathbf{R}$ matrices contain the information about the successive normalizations carried out to keep the basis orthonormal and thus about the stretching and squeezing rates. More precisely,

$$\lambda_k = \frac{1}{T_{\text{comp}}} \sum_{j=1}^{p} \ln(R_j)_{kk}, \tag{6.3}$$

where T_{comp} is the total time interval over which the computation has been carried out.

In Fig. 6.4, we show how estimates (6.3) converge to an asymptotic value with time using the Rössler system. For this attractor, we find that with an integration time $\Delta t = 10^5$, $\lambda_1 \simeq 0.0715$, $\lambda_2 \simeq 4.177 \times 10^{-6}$, and $\lambda_3 \simeq -5.394$. The exponent λ_1 (resp., λ_3) corresponds to the stretching (resp., squeezing) direction and thus is positive (resp., negative). For

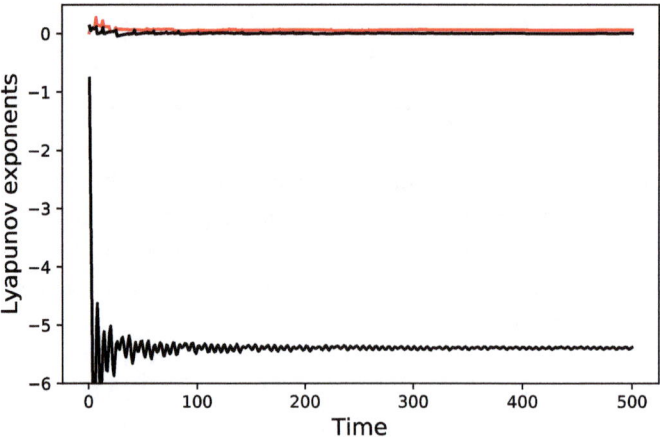

Figure 6.4: Convergence of the three Lyapunov exponents estimates λ_k given by (6.3) with computation time for the Rössler system (5.2) with $(a, b, c) = (0.2, 0.2, 5.7)$.

continuous flows, there is always a Lyapunov exponent close to zero (here λ_2) because the direction of the flow is neutral.

In the case of a recurrence system, the Lyapunov exponents are defined using one iteration step as the time unit. If we carry out the computation for the Hénon attractor of Fig 5.17, then we find that $\lambda_1 \simeq 0.408$ and $\lambda_2 \simeq -1.620$

In the case where the dynamics can be described or approached by a one-dimensional map $f(x)$, the equivalent of $\mathbf{L_X}$ at iterate x_i is the derivative $f'(x_i)$, and the equivalent of $\mathbf{Q}$ is the unit number 1, so that we have

$$\lambda = \frac{1}{p} \sum_1^p \ln|f'(x_i)|. \tag{6.4}$$

For example, (Hudson and Mankin, 1981) computed an analytical fit of the first return map shown in Fig. 5.20(c) to estimate the largest Lyapunov exponent. Applying Eq. (6.4) to their analytical fit, they found that $\lambda \simeq 0.62$.

6.1.2 Fractal dimension

It seems quite obvious to us that a line has a dimension of 1, a surface has a dimension of 2, a volume a dimension of 3, etc. Accordingly, distances are measured in meters, surfaces in square meters, and volumes in cubic meters, indicating that these different quantities scale differently with the unit chosen. In these simple examples, the dimension is equal to the number of independent coordinates that must be used to visit all the points belonging to the object. However, this definition assumes that the object is continuous in different directions.

We saw in Section 5.3.3 that in the direction where squeezing takes place, a strange attractor displays infinitely many disjoint layers stacked above each other through a recursive process, forming a self-similar structure. Such a structure is known as a *fractal*. Here we detail this concept and explain how the fractal nature of strange attractor can be characterized through a generalization of the notion of dimension, which takes non-integer values when applied to a fractal object.

6.1.2.a The box counting dimension

If we cut an object in 1, 2, 3, etc. along each of its independent directions, then the number of pieces scales differently depending on the dimension of the object (Fig. 6.5).

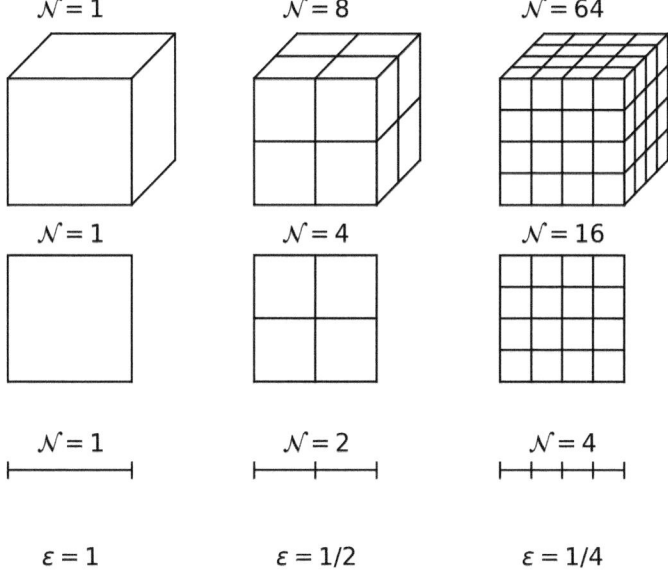

Figure 6.5: The dimension of usual geometrical objects can be defined by the scaling of the number $\mathcal{N}$ of patches of size ϵ needed to cover them. Here we show how this number evolves when we cut each side in 1, 2, or 4 for the unit segment, square, and cube.

We may thus ask how many small cubes (or squares or segments, etc.) of diameter ϵ it takes to cover a set of points in an n-dimensional space and how this number $\mathcal{N}(\epsilon)$ scales with ϵ. For usual objects (lines, surfaces, and volumes), we have $\mathcal{N}(\epsilon) \sim (1/\epsilon)^D$, where D is the usual notion of dimension (Fig. 6.5).

Accordingly, we define the box-counting dimension as

$$D_0 = \lim_{\epsilon \to 0} \frac{\ln(\mathcal{N}(\epsilon))}{\ln(1/\epsilon)}. \tag{6.5}$$

Note that the box-counting dimension of an object measures its intrinsic dimensionality independently of the dimension of a space in which we may embed it. Imagine a surface of area S in a three-dimensional space. Only one layer of cubes of diameter ϵ suffices to cover the infinitely thin surface, and there will be $\mathcal{N}(\epsilon) = S/\epsilon^2$ of them, so that $D_0 = 2$. Similar calculations are done in Table 6.1 for different types of objects.

Table 6.1: Determination of the box-counting dimension for usual geometric objects.

Object type	Extension	$\mathcal{N}(\epsilon)$	D_0
Point	0	1	0
Line segment	L	L/ϵ	1
Area	S	S/ϵ^2	2
Volume	V	V/ϵ^3	3

6.1.2.b Box-counting dimension of sample fractal sets

Now let us apply Eq. (6.5) to some classical fractal sets obtained by a recursive process where features at a given scale are defined in terms of features at a larger scale.

6.1.2.b-i Cantor ternary set

Let us take the unit segment, divide it into three, and remove the middle third. This leaves us with two segments of length $\frac{1}{3}$. This is the first step of the process. Now iterate the process, removing at each stage the middle third of any segment left so far (Fig. 6.6). The set obtained when the process is iterated an infinite number of time is called the *Cantor ternary set*.

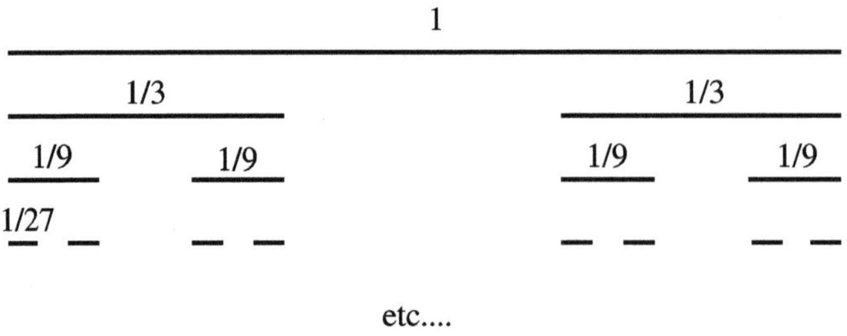

Figure 6.6: Construction of the Cantor ternary set.

If L_n denotes the total length of the set after step n, then $L_p = \frac{2}{3}L_{p-1}$, and $\lim_{p\to\infty} L_p = 0$. At stage p, we have $\mathcal{N} = 2^p$ segments, each of size $\epsilon_p = (1/3)^p$. Therefore, applying the box-counting dimension given by (6.5), we have

$$D_0 = \frac{\ln \mathcal{N}(\epsilon_p)}{\ln(1/\epsilon_p)} = \frac{\ln 2^p}{\ln 3^p} = \frac{\ln 2}{\ln 3} \simeq 0.63092\ldots,$$

which is a noninteger number. Although the Cantor ternary set has length 0, it is more than a collection of points, which have dimension 0 when isolated. It is between a point and a line.

6.1.2.b-ii Koch snowflake

Here we start from the equilateral triangle with unit sides and replace the middle third of each side by two segments of the same length (Fig. 6.7). The length now evolves like $L_p = (\frac{4}{3})^p \to \infty$.

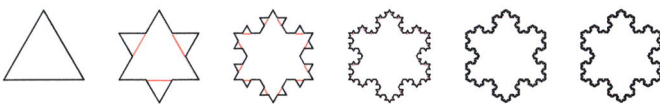

Figure 6.7: Construction of the Koch snowflake. At each stage the curve from the current (resp., previous) stage is shown in black (resp., red).

At stage p, we have $\epsilon_p = (1/3)^p$ and $\mathcal{N}(\epsilon_p) = 4^p$, so that

$$D_0 = \frac{\ln 4}{\ln 3} \simeq 1.26185\ldots.$$

Now the dimension satisfies $1 < D_0 < 2$. The Koch snowflake is intermediate between a line and a surface. Its length is infinite, and its area is zero.

6.1.2.c Box-counting dimension of the Hénon attractor

In practice, formula (6.5) cannot be used directly to measure the dimension of strange attractors because of its slow convergence. Instead, we do a log-log plot of $\mathcal{N}(\epsilon)$. If there is a fractal scaling law, then a straight line should appear, and the slope of this line provides an estimate of the fractal dimension. This amounts to defining the box-counting dimension as

$$D_0 = \lim_{\epsilon \to 0} \frac{d \ln \mathcal{N}(\epsilon)}{d \ln 1/\epsilon}. \tag{6.6}$$

Applying this method to the Hénon attractor of Fig. 5.17b, we obtain the estimate $D_0 \simeq 1.253$, showing clearly that this attractor has a fractal structure.

6.1.2.d The correlation dimension

When we analyze chaotic attractors generated by mathematical systems, we may generate as many noiseless points as we like. This is quite different when we want to characterize chaotic signals coming from experiments, where the amount of data is limited, typically, a finite sequence of points $\mathbf{X}_1, \mathbf{X}_2, \ldots, \mathbf{X}_N$, and where they may be contaminated by noise.

It has been found that the box-counting has quite poor statistical properties when dealing with experimental data. This motivated Grassberger and Procaccia (1983) to propose an alternative definition of fractal dimension, which turned out to be both more practical and efficient. In their approach, we compute the quantity

$$C(\epsilon) = \frac{1}{N^2} \sum_{i,j=1}^{N} H(\epsilon - \|\mathbf{X}_i - \mathbf{X}_j\|), \tag{6.7}$$

where H is the Heaviside function ($H(x) = 1$ if $x \geq 0$ and $H(x) = 0$ otherwise). The quantity $C(\epsilon)$ has a simple interpretation: it is the probability that the distance of two points taken at random in the strange attractor is smaller than ϵ. The behavior of this function is related with how the points are organized in space. The scaling law $C(\epsilon) \sim \epsilon^{D_2}$ defines the correlation dimension D_2, which can thus be obtained as

$$D_2 = \lim_{\epsilon \to 0} \frac{d \ln C(\epsilon)}{d \ln \epsilon}, \tag{6.8}$$

where the correlation dimension has been directly expressed in terms of a slope.

The plot of $\ln C(\epsilon)$ vs. $\ln \epsilon$ does not generally show a straight line behavior over the entire range of ϵ. Indeed, C saturates at 1 when ϵ is of the order of diameter, and conversely, there is a lack of statistics for small values of ϵ.

The computation of the fractal dimension D_2 of the Hénon attractor is shown in Fig. 6.8b for the same parameter values as in Fig. 6.8a. We can see that the two fractal dimension estimates slightly differ.

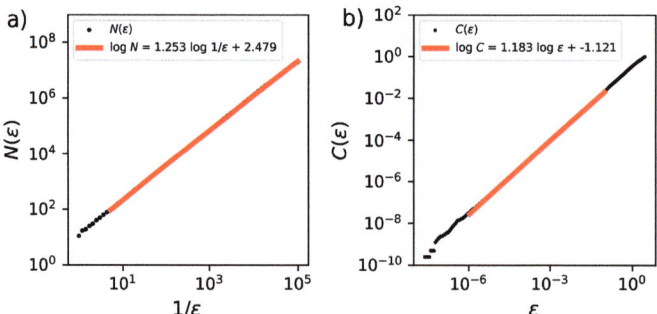

Figure 6.8: Computation of the box-counting and correlation dimensions for the Hénon map, showing that (a) $D_0 \approx 1.253$, (b) $D_2 \approx 1.183$.

6.1.2.e The Lyapunov dimension

Interestingly, there is relation between the fractal dimension of an attractor and its Lyapunov spectrum. Assume that a three-dimensional attractor with Lyapunov exponents $\lambda_1 > \lambda_2 = 0 > \lambda_3$ is continuous along the unstable and flow direction and is fractal along the stable direction with a partial dimension of η. Thus the total fractal dimension is $D = 2 + \eta$. If we assume a cover of the attractor by cubes of size ϵ, then we need

$$N_0 = K\left(\frac{1}{\epsilon}\right)\left(\frac{1}{\epsilon}\right)\left(\frac{1}{\epsilon}\right)^{\eta}$$

boxes to cover the attractor. Now if we compute the images of these cubes by the flow over a time δt, then the three axes of the cubes will be multiplied by $e^{\lambda_1 \delta t}$ (stretching), 1 (flow direction), and $e^{\lambda_3 \delta t}$ (squeezing). To cover the attractor with these deformed cubes, we would need

$$N_0' = K\left(\frac{1}{\epsilon e^{\lambda_1 \delta t}}\right)\left(\frac{1}{\epsilon}\right)\left(\frac{1}{\epsilon e^{\lambda_3 \delta t}}\right)^{\eta} = N_0 \frac{1}{e^{(\lambda_1 + \eta \lambda_3)\delta t}}$$

of them. Since the images of cubes covering the attractor also cover it, we must have $N_0 = N_0'$. Thus we conclude that $\eta = -\lambda_1/\lambda_3$, and thus

$$D_{KY} = 2 + \frac{|\lambda_1|}{|\lambda_3|},$$

an estimate known as the Kaplan–Yorke dimension, after the authors that proposed it (Kaplan and Yorke, 1979).

More generally, take j the largest integer such that $\sum_{i=1}^{j} \lambda_i \geq 0$. Then

$$D_{KY} = j + \frac{\sum_{i=1}^{j} \lambda_i}{|\lambda_{j+1}|}. \tag{6.9}$$

For the Hénon attractor, we find that $D_{KY} \simeq 1.252$ with $D_0 > D_{KY} > D_2$.

6.1.2.f Generalized dimensions

The fractal dimensions D_2 and D_0 are in fact two members of an infinite series of dimensions called the Renyi, or generalized, dimensions (Grassberger, 1983), and defined as

$$D_q = \lim_{\epsilon \to 0} \frac{\frac{1}{q-1} \ln(\sum_i p_i^q)}{\ln \epsilon}, \tag{6.10}$$

where p_i is the fraction of the attractor in box i. When $q \to 1$, D_q tends to the information dimension $D_1 = \sum_i p_i \ln p_i / \ln \epsilon$, which measures how the Shannon entropy of the distribution scales with ϵ (Rényi, 1959). D_0 and D_2 are the box-counting and correlation dimensions discussed above. The generalized dimension D_q is a decreasing function of

q because denser regions of the attractor have more weight for large values of q, thus modifying the scaling. This explains why $D_2 \leq D_1 \leq D_0$.

It is conjectured that the Kaplan–Yorke dimension is equal to D_1, which is the dimension for which the information gained by stretching due to amplifying small differences is equal to the information that is lost by squeezing.

6.1.3 Unraveling the entanglement of unstable periodic orbits

In Section 5.3.5.b, using symbolic dynamics, we established that a chaotic set contains a dense infinity of unstable periodic orbits. When we examine attentively a chaotic signal, we indeed detect many episodes of periodic behavior where the signal almost repeats itself over a short time interval (Fig. 6.9a). Each of these episodes corresponds to the system visiting a close neighborhood of an unstable periodic orbit. Low-period orbits, such as the period-1 orbit that appears clearly on the second row of Fig. 6.9a, are less unstable and are thus more frequently visited.

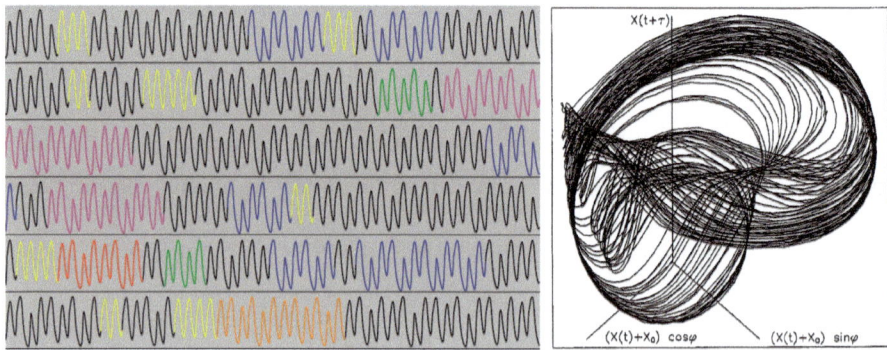

Figure 6.9: (a) Short segments of periodic behavior with period up to 5 detected in the signal of Fig. 5.18. The color indicates the period. (b) Closed orbits of period up to 12 detected in the same signal. Reprinted from (Lefranc and Glorieux, 1993).

In Fig. 6.9b the closed trajectories associated with periodic episodes of period up to 12 have been plotted, and Fig. 5.18b shows that together they provide an excellent approximation of the strange attractor. Closed periodic orbits, being finite in length, are easier to characterize robustly than long chaotic trajectories.

For three-dimensional chaotic systems, the entanglement of periodic orbits can be characterized using knot theory. Knot theory provides us with topological invariants that quantify how closed curves are intertwined and do not vary when they are deformed continuously without making them cross themselves. For example, the period-1 and period-4 orbits in Fig. 6.10a can be deformed to the configuration of Fig. 6.10b, where we can see that their linking number is 2 and the period-1 orbit is not knotted, whereas

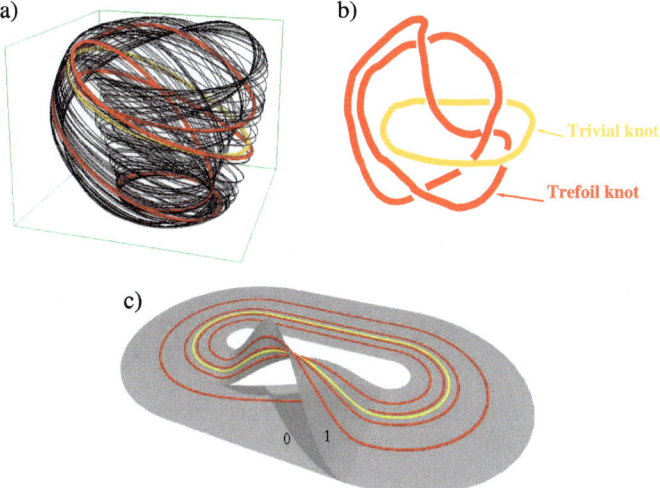

Figure 6.10: (a) Period-1 and period-4 orbits embedded in a strange attractor. These orbits can be deformed to the configuration of (b), where we see that the period-4 orbit makes two turns around the period-1 orbit. Mathematically speaking, their linking number is 2. The period-1 orbit is not knotted, whereas the period-4 orbit is a trefoil knot, the simplest of all knots. (c) A branched manifold carries the two orbits in the same configuration as in b). It shows that the two orbits have been entangled by a horseshoe-like mechanism.

the period-4 orbit is a trefoil knot, which is the simplest nontrivial knot. This is consistent with the attractor being generated by a horseshoe-like mechanism (Fig. 6.10c).

Using this approach, the knot types of all orbits and linking numbers of all pairs of orbits in the attractor can be computed. From these invariants it is possible to compute a branched manifold such as in Fig. 6.10c, called a *template*, that can carry all orbits up to continuous deformation and thus summarizes the global organization of the attractor (Fig. 6.11). This provides a robust characterization of the stretching and folding mechanisms generating chaos. For further information, see (Gilmore and Lefranc, 2011).

6.2 Transition toward chaos in the logistic map

In Section 5.3, we analyzed the structure of strange attractors and the mechanisms generating them. Section 6.1 provided us with the quantitative tools needed to characterize them. However, we still have to understand how a system evolves from regular behavior to aperiodicity as a parameter varies.

Does the complexity appears once and for all, or is it created gradually? What elements is it made of? In this section, we delve into the details of how chaos is born. In doing so, we will find that deterministic chaos is not an "amorphous" dynamical state, but that it is actually exquisitely organized and structured.

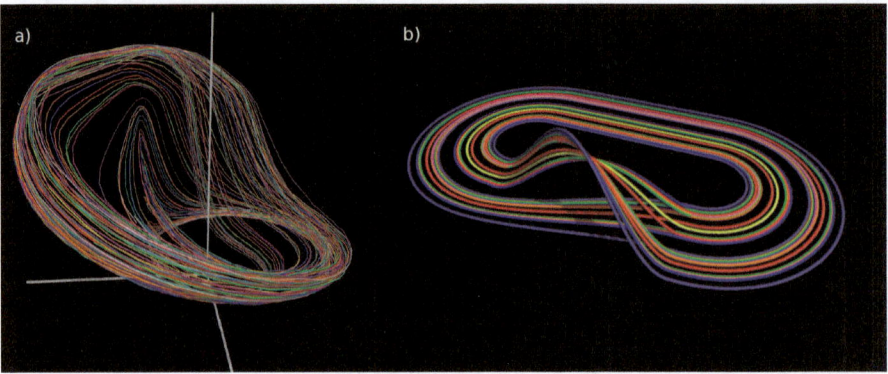

Figure 6.11: By computing the topological invariants of (a) a set of unstable periodic orbits embedded in a chaotic attractor, it is possible to compute (b) a branched manifold that carries a set of orbits with the same invariants and characterizes the geometrical mechanisms generating chaos, here a simple horseshoe.

In this introductory book, we will restrict ourselves to one type of transition, the so-called *period-doubling* or *subharmonic cascade*, a transition that is ubiquitous in low-dimensional ordinary differential equations. Indeed, the mechanisms leading to chaotic behavior are quite generic and depend little on the specificities of the system studied. Many distinct physical systems, described by very different equations, display exactly the same universal scenarios when they plunge into chaos. This makes it possible to think about chaos theory, rather than about the theory of the different chaotic behaviors of systems A or B. As a result, there are only a few well-identified routes to chaos. The period-doubling cascade is the most common one. The transition to chaos via quasi-periodicity, also known as the Ruelle–Takens scenario (Ruelle and Takens, 1971), is another one.

Moreover, we will restrict ourselves to the study of a simple one-dimensional recurrence system, the logistic map. However, this will barely limit our understanding of the period-doubling cascade, as such systems are faithful representatives of strongly dissipative flows with a single unstable direction (Fig. 5.10) and reproduce to a large extent the features of less dissipative systems. This covers many relevant systems in nature, in particular, when chaos first appears from ordered motion with a single unstable direction.

6.2.1 Introduction

In this section, we study the dynamics of what is possibly the simplest dynamical system, the logistic map recurrence

$$x_{n+1} = f(x_n) = rx_n(1 - x_n)$$

with $r > 0$.

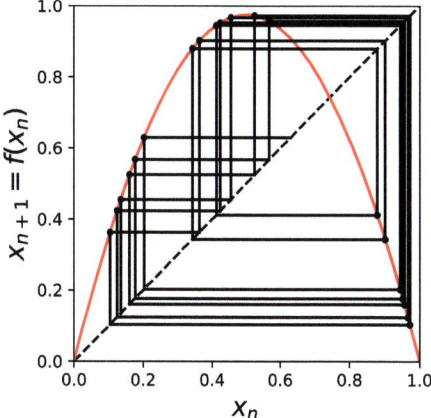

Figure 6.12: The graph of the logistic map for $r = 3.9$, shown with the diagonal $y = x$ and a part of a chaotic orbit drawn as explained in Section 1.4.2.

As Fig. 6.12 shows, a key feature of the logistic map is that its graph is nonmonotonous: it has an extremum at the *critical* point $x_c = \frac{1}{2}$, whose image is $f(x_c) = \frac{r}{4}$. As a consequence, it is noninvertible: two points located symmetrically around the critical point have the same image. This ingredient produces the folding process that we have discussed in Section 5.3.2. Since two distinct points are collapsed onto each other in exactly one step, this corresponds to the case of infinite dissipation. Such a function with a single maximum is called a *unimodal* map.

For a one-dimensional system that displays continuous stretching, the noninvertibility is essential to maintain the dynamics within a bounded region. Indeed, there is no other direction in which squeezing and folding can take place to counteract the expansion.

As with the horseshoe map (Section 5.3.5), a symbolic coding of the orbits of the logistic map is possible. To this aim, we keep track of which side of the critical point x_c a point and its successive iterates fall:

$$\Sigma(x) = s_0 s_1 s_2 \ldots s_i \ldots, \quad s_i = \begin{cases} 0 & \text{if } f^i(x) \leq x_c, \\ 1 & \text{if } f^i(x) > x_c. \end{cases} \tag{6.11}$$

For the same reasons we invoked for the horseshoe, a given symbolic sequence is associated with one and only one point, which is thus labeled uniquely (the logistic map is a horseshoe with infinite squeezing rate). This symbolic coding will be useful when we will need to enumerate the periodic points of the logistic map.

Moreover, it is useful to have an order relation on sequences that preserves the ordering of points along the interval so that

$$\Sigma(x) < \Sigma(x') \quad \Leftrightarrow \quad x < x'. \tag{6.12}$$

If two sequences begin by 0 and 1, respectively, then it is easy to order them since we know that the associated points are left and right of x_c: $0 \ldots \prec 1 \ldots$. If we now consider two sequences with i identical leading symbols, such as

$$\Sigma(x) = 0.s_0 s_1 s_2 \ldots s_{i-1} s_i \ldots,$$
$$\Sigma(x') = 0.s_0 s_1 s_2 \ldots s_{i-1} s_i' \ldots, \tag{6.13}$$

then we can order $\sigma^i \Sigma(x)$ and $\sigma^i \Sigma(x')$ by shifting them to the left until the leading digit differs. Then we know how $f^i(x)$ and $f^i(x')$ are located relative to each other. To deduce the relative position of x and x', we just have to take into account how many times we went through the orientation-reversing branch of f (Fig. 6.12), which inverts the order of the two iterates. This amounts to count the number of 1s in the common part of the two sequences:

$$\Sigma(x) \prec \Sigma(x') \quad \Leftrightarrow \quad \begin{cases} s_i < s_i' & \text{and} \quad \Pi_{k=0}^{i-1}(-1)^{s_k} = 1, \\ s_i > s_i' & \text{and} \quad \Pi_{k=0}^{i-1}(-1)^{s_k} = -1. \end{cases} \tag{6.14}$$

Then property (6.12) is satisfied.

To summarize, we can order two sequences according to how the corresponding points are ordered along the real line. We shift them simultaneously until their leading digit differ, allowing us to order them. Having kept track of how many orientation reversals we have performed, we can order the initial sequences. For example (check!),

$$011 \ldots \prec 101 \ldots$$
$$110 \ldots \prec 101 \ldots$$
$$0110 \ldots \prec 0111 \ldots$$
$$10111 \ldots \prec 10110 \ldots$$

6.2.2 Fixed points of the logistic map

As $f([0,1]) = [0, \frac{r}{4}]$, the interval $[0,1]$ is a trapping region for $r \in [0,4]$: any orbit that starts in it remains in it, and we can then restrict our study to this interval. When $r > 4$, we can check that any orbit diverges to $-\infty$; consequently, we will not consider this case.

The fixed points of the logistic map are given by the equation $x^* = rx^*(1 - x^*)$. It has two obvious solutions for all values of r:

$$x_1^* = 0, \quad x_2^* = 1 - \frac{1}{r}. \tag{6.15}$$

However, $x_2^* \in [0,1]$ only when $1 \le r$. The two fixed points in (6.15) collide and exchange their relative positions when $r = 1$ as shown in Fig. 6.13.

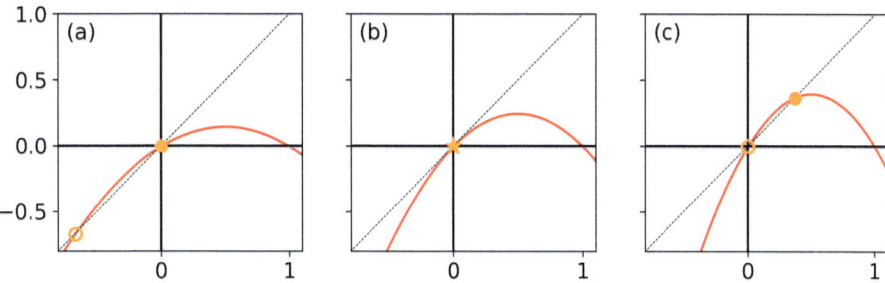

Figure 6.13: Graph of the logistic map for (a) $r = 0.6$, (b) $r = 1$, (c) $r = 1.6$.

To determine the stability of these fixed points, we compute their multipliers: $\mu_i = f'(x_i^*) = r(1 - 2x_i^*)$ (Section 1.4.2), giving $\mu_1 = r$ and $\mu_2 = 2 - r$. As a fixed point is stable when $|\mu_i| \leq 1$, $x_1^* = 0$ is stable for $0 \leq r \leq 1$, and the two points exchange their stability at $r = 1$, where the graph of $f(x)$ is tangent to the diagonal in 0 (consistent with $\mu_{1,2} = 1$). We recognize the transcritical bifurcation studied in Section 3.2.4.b.

Interestingly, the fixed point $x_2^* = 1 - \frac{1}{r}$ is stable only for $1 \leq r \leq 3$, which implies that for $r \geq 3$, both fixed points are unstable. As a matter of fact, $\mu_2 = -1$ at $r = 3$, indicating that small deviations from the fixed point alternate between the two sides of the fixed point. This is typical of the period-doubling bifurcation we have studied in Section 4.2.2.b, as we will see below.

6.2.3 The subharmonic cascade

6.2.3.a The period-2 orbit

For r slightly above 3, the permanent regime consists of oscillations between two values, as illustrated graphically in Fig. 6.14. Note that the fixed point $x_2^* = 1 - \frac{1}{r}$ is still present, as shows the intersection of the graph of f with the diagonal, but it is now unstable.

To understand how the fixed point (or period-one orbit) destabilizes to give birth to this period-two orbit, we note that the latter is in fact a fixed point of the doubly iterated map $f^2 = f \circ f$. Indeed, if the dynamics oscillate between $\tilde{x}_2 = f(\tilde{x}_1)$ and $\tilde{x}_1 = f(\tilde{x}_2)$, then we have $\tilde{x}_1 = f(f(\tilde{x}_1)) = g(\tilde{x}_1)$ and $\tilde{x}_2 = f(f(\tilde{x}_2)) = g(\tilde{x}_2)$, indicating that $\tilde{x}_1$ and $\tilde{x}_2$ are both fixed points of f^2.

– For $r \leq 3$ (Fig. 6.15a), the graph $y = f^2(x)$ has two intersection points with the diagonal $y = x$. These correspond to the fixed points of f (Eq. (6.15)), which are necessarily also fixed points of $f^2 = f \circ f$. The fixed point at $x_2^* = 1 - \frac{1}{r}$ is stable since the slope of the graph is smaller than 1, crossing the diagonal from above to below.

– When $r = 3$ (Fig. 6.15b), the graph $y = f^2(x)$ is tangent to the diagonal at $x_2^* = 1 - \frac{1}{r}$, indicating a structurally unstable and degenerate situation. Since the graph crosses the diagonal rather than being on one side only, we know that there are at least

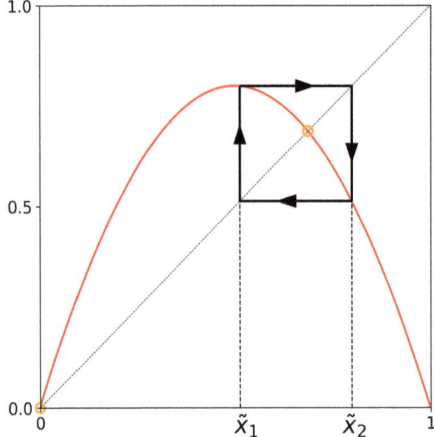

Figure 6.14: Period-2 orbit of the logistic map for $r = 3.2$: iterates oscillate between two different values $\tilde{x}_1$ and $\tilde{x}_2$.

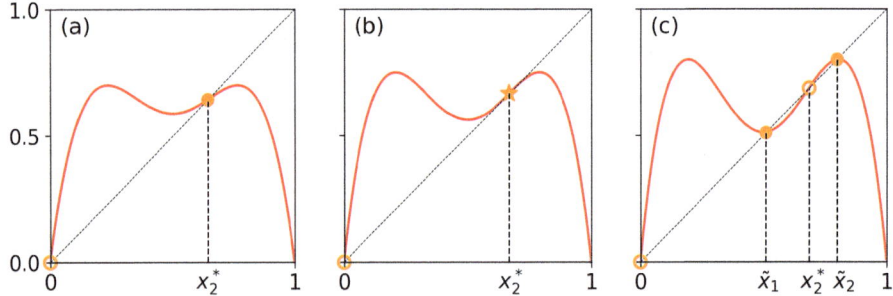

Figure 6.15: Evolution of the graph of $f^2(x) = f(f(x))$ as we go through the period-doubling bifurcation at $r = 3$: (a) $r = 2.8$, (b) $r = 3$, (c) $r = 3.2$.

three coincident fixed points. These fixed points are marginally stable (the slope of the graph is exactly one).

– When $r > 3$ (Fig. 6.15c), two fixed points appear on each side of the fixed point $x_2^* = 1 - \frac{1}{r}$. By examining the slopes of the graph at the intersection points we see that x_2^* (fixed point for both f and f^2) is unstable and that the two new fixed points are stable.

We now study this bifurcation algebraically. First, we compute the expression of $f^2(x)$:

$$
\begin{aligned}
f^2(x) &= rf(x)(1 - f(x)) \\
&= r[rx(1-x)][1 - rx(1-x)] \\
&= r^2 x(1-x)[1 - rx(1-x)].
\end{aligned}
$$

Fixed points of f^2 are roots of the polynomial

$$f^2(x) - x = r^2 x(1-x)\big[1 - rx(1-x)\big] - x,$$

which can easily be factorized knowing that 0 and $1 - \frac{1}{r}$ (fixed points of f) are obvious roots. We then obtain the values of the two points of the period-2 orbit:

$$\tilde{x}_1 = \frac{1 + r - \sqrt{(r-1)^2 - 4}}{2r} \quad \text{and} \quad \tilde{x}_2 = \frac{1 + r + \sqrt{(r-1)^2 - 4}}{2r}.$$

They indeed are defined only for $r \geq 3$, since this orbit appears in the period-doubling bifurcation at $r = 3$.

To study the stability of these fixed points, we have to calculate $\frac{d}{dx} f^2(\tilde{x}_{1,2}) \equiv f^{2'}(\tilde{x}_{1,2})$ knowing that $f^{2'}(x) = f'(x) f'(f(x))$, so that

$$f^{2'}(\tilde{x}_1) = f'(\tilde{x}_1) f'(f(\tilde{x}_1)) = f'(\tilde{x}_1) f'(\tilde{x}_2) = f^{2'}(\tilde{x}_2).$$

The two proper fixed points of f^2 have the same multiplier, which is indeed the multiplier of the period-2 orbit of f that they make together and is given by

$$\tilde{\mu}_{1,2} = f^{2'}(\tilde{x}_{1,2}) = -r^2 + 2r + 4.$$

We find that this multiplier monotonously decreases from $\tilde{\mu}_{1,2} = 1$ at $r_1 = 3$ to $\tilde{\mu}_{1,2} = -1$ at $r_2 = 1 + \sqrt{6} = 3.449489\ldots$, which indicates a new period-doubling bifurcation, repeating the scenario that we have observed for the period-1 orbit.

6.2.3.b Second period-doubling bifurcation: the period-4 orbit

At $r_2 = 1 + \sqrt{6} = 3.449489\ldots$, the period-2 orbit has multiplier -1, indicating that it destabilizes through a period-doubling bifurcation. As a fixed point of $f^4 = f^2 \circ f^2$, it has thus a multiplier $1 = (-1)^2$, which is the signature of a structurally unstable situation. A period-4 orbit is born, which coincides with the destabilized period-2 orbit at the bifurcation.

The analysis that we carried out for the initial period doubling bifurcation around $r_1 = 3$ can be done again around $r = 3.449489\ldots$ studying the map $f^4 = f^2 \circ f^2$. In particular, the four new fixed points of f^4 can be identified if we plot f^4 side by side with f^2 for $r = 3.54 > 3.449489\ldots$ (Fig. 6.16(a,b)).

If r is further increased, the period-4 orbit destabilizes at $r = 3.5440904\ldots$ to give birth to a period-8 orbit. The graph of the function f^8 is plotted in Fig. 6.16(c), where we can see that the function has 16 intersections with the diagonal $y = x$. The 8 new fixed points are stable and correspond to the period 8 orbit.

The successive period-doubling bifurcations discussed above are in fact the first stages of an infinite series of bifurcations where the same scenario repeats itself again and again: an orbit of period 2^{n-1} destabilizes (its multiplier crosses -1), giving birth to an orbit of period 2^n. The multiplier of the latter is initially 1 and decreases gradually

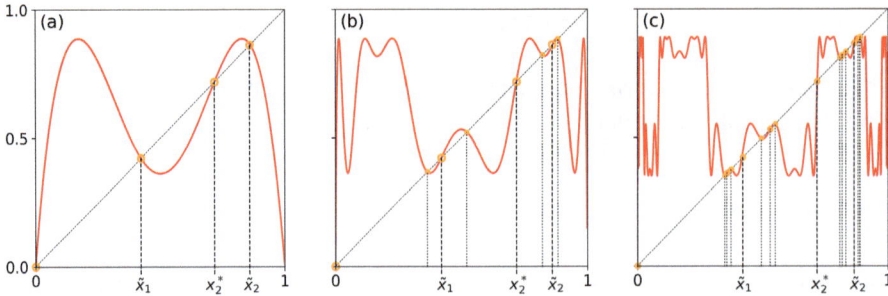

Figure 6.16: Plot of (a) f^2 and (b) f^4 for $r = 3.54$. The four fixed points of f^4 that are not fixed points of f^2 correspond to a period-4 orbit born in the period-doubling cascade at $r = 3.449489\ldots$. (c) Plot of f^8 for $r = 3.56$.

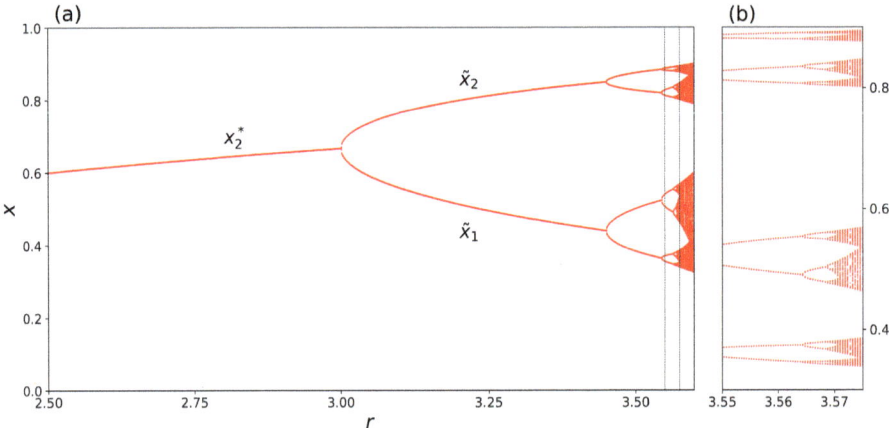

Figure 6.17: (a) Bifurcation diagram of the logistic map for $r \in [2.5, 3.6]$, showing the first stages of the period-doubling cascade. (b) Enlargement for $r \in [3.55, 3.75]$, showing the successive bifurcations.

to -1, where the next period-doubling bifurcation gives birth to an orbit of period 2^{n+1}. Between 1 and -1, there is a parameter where the multiplier is 0: the periodic orbit is then said to be *superstable*, as perturbations converge to zero extraordinarily fast.

This scenario is summarized in Fig. 6.17(a). Figure 6.17(b) is an enlargement for $r \in [3.55, 3.75]$, showing the period-8 orbit and the subsequent bifurcations.

We observe that the domain of stability of each 2^n-orbit decreases rapidly with n. Table 6.2 gives the parameter values at which the first period-doubling bifurcations occur. We can see that they are indeed closer and closer as n increases and that they in fact follow a geometric sequence such that the ratio $(r_{i-1} - r_i)/(r_i - r_{i+1})$ is approximately constant. Consequently, an infinite number of bifurcations occur over a finite interval of r, terminating with an orbit of period 2^∞ for $r = r_\infty$. At this point the motion is no longer periodic, since the period is infinite. We are entering chaos.

Table 6.2: Values of the control parameter r for which a period-doubling bifurcation is observed in the initial period-doubling cascade. The quantities tabulated in the last column indicate that these values follow a geometric sequence of ratio $\delta = 4.6692013\ldots$.

i	bifurcation	r_i	$\delta_i = (r_{i-1} - r_i)/(r_i - r_{i+1})$
1	$1 \rightarrow 2$	3.0	
2	$2 \rightarrow 4$	3.4494897	4.7514462
3	$4 \rightarrow 8$	3.5440904	4.656251
4	$8 \rightarrow 16$	3.5644073	4.6682422
5	$16 \rightarrow 32$	3.5687594	4.6687395
6	$32 \rightarrow 64$	3.5696916	4.6691322
7	$64 \rightarrow 128$	3.5698913	4.669183
8	$128 \rightarrow 256$	3.5699340	4.6691981
9	$256 \rightarrow 512$	3.5699432	4.6692008
10	$512 \rightarrow 1024$	3.5699451	4.6692013
11	$1024 \rightarrow 2048$	3.5699456	
...	...	...	
∞	$2^{\infty} \rightarrow 2^{\infty}$	3.569945672	

We will see later that the convergence rate of the sequence series of bifurcation parameter values ($\delta = 4.6692013\ldots$) is universal. This exact value is found for any map of the interval into itself with a quadratic maximum. It is also observed for period-doubling cascades in higher-dimensional maps or ODE systems. Indeed, period-doubling bifurcations in such systems are essentially a one-dimensional phenomenon occurring along the direction in which the orbit is destabilizing (see, e. g., Section 4.2.2.b).

6.2.4 Beyond the initial period-doubling cascade

6.2.4.a Bifurcation diagram and Lyapunov exponents

For $r \geq r_{\infty}$, we observe many regimes where the recurrence is no longer periodic, with the iterates taking infinitely many different values. To characterize the nature of the dynamics in this range, we can compute the Lyapunov exponent λ such as defined by Eq. (6.4). The result is plotted in Fig. 6.18(b) together with the complete bifurcation diagram (Fig. 6.18(a)).

Coming back to $r < r_{\infty}$, we see that the Lyapunov exponent is negative over the whole parameter range as expected for stable periodic regimes, only taking the value 0 at each period-doubling bifurcation (dashed lines in Figs. 6.18(a,b)). It is even infinitely negative at the superstable points. Beyond r_{∞}, λ is positive for most values of r, identifying the presence of chaotic behavior. However, we also find negative values of λ for different intervals in r, revealing the existence of many windows of periodic behavior beyond r_{∞}. In fact, the largest periodic windows are clearly discernible in the bifurcation diagram of Fig. 6.18(a), where they appear as white stripes. We will look more closely at these periodic windows in Section 6.2.4.b.

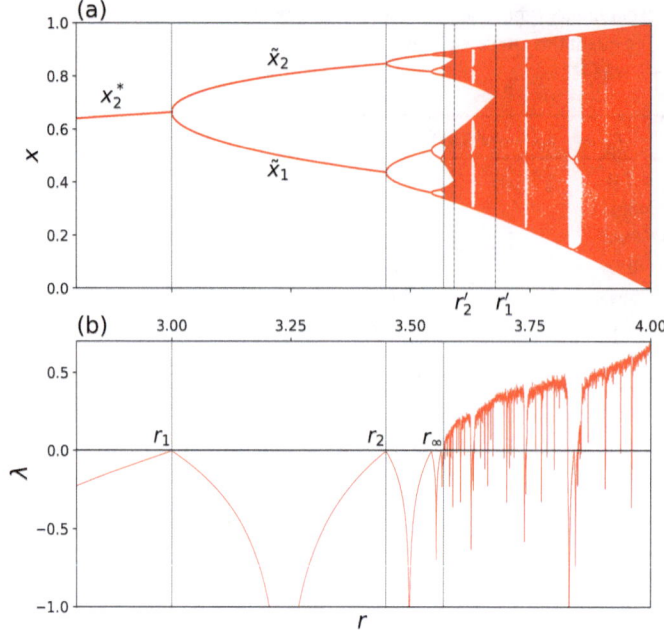

Figure 6.18: (a) Bifurcation diagram of the logistic map for $r \in [2.8, 4]$. Dash-dotted lines indicate the first two steps of the inverse cascade. (b) Lyapunov exponents as functions of parameter r. Dashed lines common to the two graphs: values of r_1, r_2, and r_∞ of Table 6.2.

At the accumulation point of the period-doubling cascade, the iterates are contained in a Cantor set. This is related to the existence of a scale invariance in the period-2^∞ orbit, as we will discuss in Section 6.2.5. Immediately beyond the transition to chaos at $r = r_\infty$, the iterates are spread over infinitely many intervals, each of them being densely filled. Then these chaotic bands merge progressively, dividing the number of bands by two at each merging, until there is only one band left at $r = r_1' \approx 3.678$. This band-merging cascade is a mirror of the period-doubling cascade, with r_i' denoting the parameter value where 2^i bands merge into 2^{i-1} ones and is known as the *inverse cascade*. The band-merging events also follow a geometric sequence with the same universal exponent as the direct sequence, that is,

$$\frac{r_{n-1}' - r_n'}{r_n' - r_{n+1}'} \to \delta,$$

and they join the direct cascade at $r_\infty' = r_\infty$, where both accumulate on the same Cantor set.

Moreover, the total length of the chaotic intervals increases with r until the whole interval $[0, 1]$ is visited for $r = 4$. At this point, the chaotic attractor collides with the unstable fixed point (or period-1 orbit) that we studied in Section 6.2.2, which constitutes the basin boundary of the attractor. Beyond $r = 4$, most iterates escape to $-\infty$, except for a Cantor set of measure 0. The invariant set that contains all the orbits created for $r \leq 4$ is no longer an attractor, but remains as a *repulsor*.

6.2.4.b Periodic windows

Over the parameter range $[r_\infty, 4]$, we find an infinity of periodic windows such as those displayed in Fig. 6.19, which shows a period-5 window (for $3.738 \leq r \leq 3.745$) and the unique period-3 window of the diagram (for $3.828 \leq r \leq 3.857$). The periodicity is indicated by the finite number of values taken in the bifurcation diagram and also by the negative Lyapunov exponent.

The scenario from the beginning to the end of the window is identical in all these windows and is as follows. First, the periodic behavior emerges abruptly from an irregular regime. We will see that this corresponds to a saddle-node bifurcation where a new stable periodic orbit appears (together with an unstable one, as we can recall) and captures neighboring orbits on its whole domain of stability. As r increases, a period-doubling cascade is initiated, which brings back the system to chaos, but the bifurcation diagram is still partitioned into p intervals, where p is the fundamental period of the window. Then, suddenly, the dynamics explores the whole accessible interval again. This event is known as a *crisis* and is well understood: it occurs when the chaotic attractor born from the stable periodic orbit at the beginning of the window collides with the unstable periodic orbit that appeared with it.

In fact, the whole sequence of events between the beginning of the window and the crisis recapitulates the whole bifurcation diagram of the logistic map, except that it is duplicated over the p branches. In particular, a three-branch window can be discerned inside the middle branch of the period-3 window (Fig. 6.19b), actually located within a period-9 window. Inside this "period-three window", there is another one, and so on ad libitum. As if we had not had enough complexity in our journey into chaos, it turns out that the bifurcation diagram of the logistic map is itself a fractal. This fractal structure

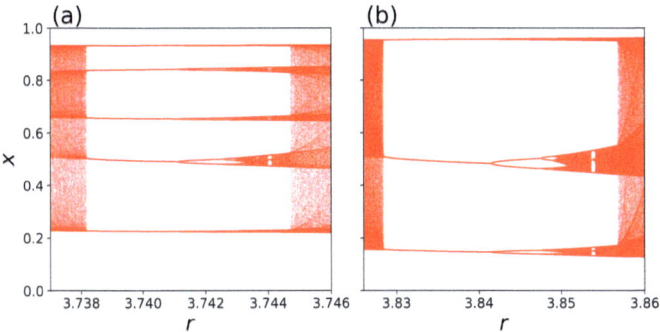

Figure 6.19: Periodic windows of period (a) 5 and (b) 3 embedded in the bifurcation diagram of Fig. 6.18.

can be understood in terms of the symbolic dynamics we introduced in Section 5.3.5 (see (Gilmore and Lefranc, 2011) for details).

In Fig. 6.18a, we discern only a finite number of periodic windows because of the finite resolution. In the Lyapunov exponent plot (Fig. 6.18b), we detect a few more, but still a finite number. However, there are infinitely many more. Figure 6.20 shows the bifurcation diagram of the logistic map beyond the accumulation point with the parameter values where a periodic orbit of period equal to or lower than 8 is superstable. Each corresponds to a different periodic window. We can see that many windows cannot even be discerned with a naked eye except for the lowest-period ones. Orbits located at the beginning of a window appear through a saddle-node bifurcation, and the other through a period doubling bifurcation (see the period-4 and period-8 orbits near $r = 3.96$). In the logistic map, periodic orbits born for some value of r exist all the way through $r = 4$, implying the presence of more and more periodic orbits as we proceed to fully developed chaos at $r = 4$.

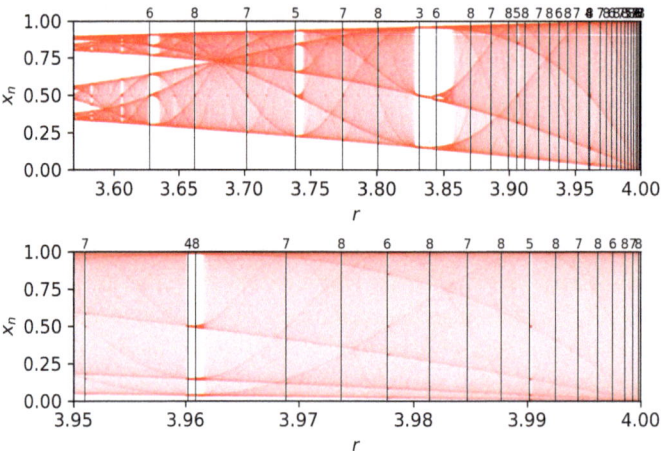

Figure 6.20: Bifurcation diagram of the logistic map between $r = r_\infty$ and $r = 4$, where the parameter values where a periodic orbit is superstable have been indicated with a vertical line, with the period indicated at the top. The bottom part is an enlargement of the diagram for the interval $r \in [3.95, 4.0]$.

We saw in Section 6.2.1 that any point can be associated with an infinite symbol sequence (Eq. (6.11)). This is in particular true for periodic points, which are associated with periodic sequences of 0s and 1s. This implies that there is a countable infinity of periodic windows of arbitrarily high period, deeply intertwined with the chaotic regions. For example, there are $2^8 = 256$ periodic points of period 8, but 2 of them are actually of period 1, 2 of them of period 2, and 12 of them are of period 4. There are thus $240/8 = 30$ periodic orbits of lowest period 8. Of these, one is born in the initial period-doubling cascade, one is the period-doubled orbit of the period-4 window, and the other are born in 14 different saddle-node bifurcations. More generally, there are roughly $2^p/p$ orbits of period p for arbitrarily large p. All the corresponding windows are embedded in the bifurcation diagram, whose complexity we can now appreciate.

Now that we have realized the ubiquity of periodic windows, let us study more closely the period-3 window, which is the largest one. To understand the emergence of this orbit at $r = r_{c3}$, we study the fixed points of f^3.

- Fig. 6.21(a) shows the graph of f^3 for $r < r_{c3}$. The graph has two intersections with the diagonal $y = x$: the fixed points x_1^* and x_2^* of f, which are both unstable.
- At $r = r_{c3}$ (Fig. 6.21(b)), the graph of f^3 becomes tangent to $y = x$ in three points simultaneously.
- A further increase of r leads to the emergence of 6 new fixed points of f^3 through 3 simultaneous saddle-node bifurcations (Fig. 6.21(c)). These three fixed points of f^3 form a period-3 orbit of f. The sudden appearance of this stable orbit at $r = r_{c3}$ explains why chaos is suddenly interrupted at the beginning of the corresponding periodic window.

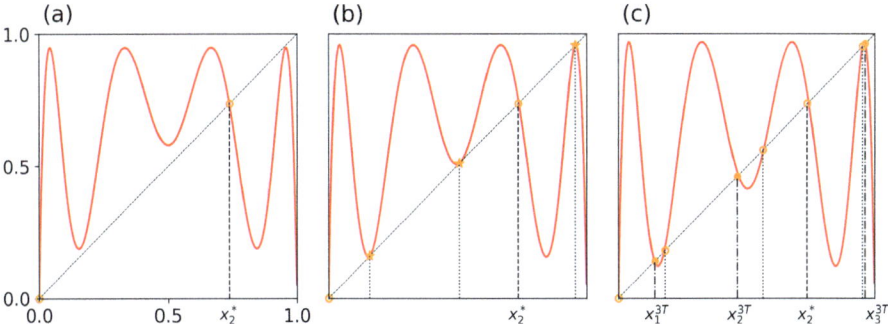

Figure 6.21: The graph of f^3 is plotted for (a) $r < r_{c3}$, where it only intersects the diagonal at the fixed points of f; (b) $r = r_{c3}$, where 6 new intersections appear through a tangency with the diagonal, indicating a saddle-node bifurcation; (c) $r > r_{c3}$, where these 6 intersections have evolved into the periodic points of a stable and unstable period-3 orbits of f.

Let us now study how the period-3 orbit becomes unstable. The slopes of the graph of f^3 at the three fixed points x_i^{3T} ($i = 1, 2, 3$) are identical as

$$f^{3\prime}(x_i^{3T}) = f'(x_i^{3T})f'(f(x_i^{3T}))f'(f^2(x_i^{3T})) = f'(x_1^{3T})f'(x_2^{3T})f'(x_3^{3T}).$$

This common value is the multiplier $\mu_3 = f^{3\prime}(x_i^{3T})$ of the orbit. It decreases with increasing r, starting from $\mu_3 = 1$ when $r = r_{c3}$. When it reaches -1, the period-3 orbit destabilizes through a period-doubling bifurcation, as can be seen in Fig. 6.19(b). A period-doubling cascade then follows leading to orbits of period 3×2^n for all n.

The same scenario is observed for all the other periodic windows (see, e. g., the 5T window in Fig. 6.19(a)). There exist an infinite number of periodic windows, each locally replicating the overall structure of the period-doubling cascade.

6.2.5 Universality

What makes the study of the bifurcation diagram of logistic map so valuable because many of its properties are *universal*: they do not depend of the details of the function. All the maps of an interval into itself that have a single quadratic maximum share the same structure of the bifurcation diagram.

One aspect of this universality is the order in which the different periodic windows appear. Another one is the structure of the period-doubling cascade and the structure of the invariant set at the accumulation point of the cascade.

6.2.5.a The universal sequence

We saw that there are infinitely many periodic windows in the bifurcation diagram of the logistic map. However, what determines whether a given periodic orbit is created before or after another one as one proceeds toward fully developed chaos at $r = 4$?

For this purpose, it is useful to label a periodic orbit with the symbol sequence associated with its rightmost periodic point, as determined by (6.11). For example, 1011 is the sequence of the rightmost point of the period-4 orbit of the initial period-doubling cascade, whose name can be obtained as follows. When the period-1 orbit experiences period doubling, its name is 1 since it is located on the orientation-reversing part of the graph. After doubling, we have a period-2 orbit 11, which becomes 10 after crossing the superstable point, where the last symbol of the sequence is flipped, as will be explained below. Through period-doubling, this orbit gives birth to a period-4 orbit 1010, which itself becomes 1011 before undergoing period-doubling. The next orbit in the cascade would be the period-8 orbit 10111010. We can see that the symbolic name of an orbit tells us its genealogy.

When a periodic orbit is superstable, the rightmost point is the image $f(x_c)$ of the critical point and is thus located at $x = \frac{r}{4}$, the largest value any iterate can take. Moreover, different periodic points cannot cross as r is changed because, as fixed points of some f^k, they remain isolated away from a bifurcation. Hence, the order of existing periodic points is preserved as r is increased.

Thus we conclude that if the rightmost point of a periodic orbit is to the right of the rightmost point of another periodic orbit, then the latter must have appeared before the former. Indeed, the more to the right the rightmost point, the larger the r when the orbit was superstable.

Orbits that appear together in the saddle-node bifurcation creating a new periodic window are initially superimposed and hence have the same symbol sequence (which is possible because the dynamics is not then chaotic). For example, the two orbits appearing at the beginning of the first period-5 window have itinerary 11011, with a even number of 1s since they have positive multiplier. The initially stable orbit then becomes superstable at some value of r, where one of its periodic point crosses the maximum, changing the side and thus the symbol. Because the rightmost point is the image of the point closest to maximum, the last symbol of the rightmost point changes when the first symbol of its preimage changes, since applying the map is equivalent to shifting the sequence to the left.

Here the stable orbit becomes 11010 at the superstable point. Since it has now a negative multiplier (indicated by the odd number of 1), it can undergo a period-doubling. The initially unstable orbit remains so and keeps its 11011 name. More generally, when the sequences of two periodic orbits differ only in the last symbol, we know that they have appeared together in the same saddle-node bifurcation. Similarly, we may identify which orbits are involved in the same period-doubling bifurcation. When we flip the last symbol of the sequence associated with a period-doubled orbit, it indeed becomes a twice repeated pattern. For example, $100101 \to 100100 = (100)^2$, indicating that 100101 is the period-doubled orbit of 100.

Putting these facts together allows us to determine the order of appearance of all orbits in the bifurcation diagram given their symbolic name, including those in the initial period-doubling cascade. To achieve this, we enumerate all possible periodic sequences of 0s and 1s up to a certain period p. For each of them, we shift it repeatedly until we find the rightmost sequence, and we build a unique list of these rightmost points. From the above discussion, we only need to consider the sequences with an odd number of 1, since the other ones are the unstable saddles. Then we order these rightmost points using the order (6.14), which gives their order of appearance. This order only depends on the fact that there is a single maximum in the map and is known as the *universal sequence*. The sequence of bifurcations up to period 6 is shown in Table 6.3.

Table 6.3: Universal sequence of periodic windows observed in unimodal maps with the symbolic names of the orbits involved, up to period 6. w_p^n denotes the nth window of period p, and $w_p^n \times 2^k$ is the kth period doubling inside w_p^n. For each window, the value of r for which the stable orbit is superstable for the logistic map is indicated, allowing us to locate the window in Fig. 6.20.

Win.	Name	r_{ss}	Win.	Name	r_{ss}	Win.	Name	r_{ss}
w_1	$^1_0$	2.0	w_3^1	10_0^1	3.832	w_6^3	10001_0^1	3.978
$w_1 \times 2^1$	10	3.236	$w_3^1 \times 2^1$	100101	3.845	w_5^3	1000_0^1	3.990
$w_1 \times 2^2$	1011	3.499	w_5^2	1001_0^1	3.906	w_6^4	10000_0^1	3.998
w_6^1	10111_0^1	3.628	w_6^2	10011_0^1	3.938			
w_5^1	1011_0^1	3.739	w_4^1	100_0^1	3.960			

This symbolic dynamical approach allows us to understand the fractal structure of the bifurcation diagram. For example, all the periodic orbits of the period-3 window have sequences made of the two symbols $X = 101$ and $Y = 100$, which are the two period-3 orbits born in the saddle-node bifurcation, in the same way as all periodic orbits have sequences made of 1 and 0. Thus there is a one-to-one correspondence between orbits with sequences of 0s and 1s and orbits with sequences of X and Y. The simplest example is the period-2 orbit 10, born in the initial period-doubling bifurcation of the period-1 orbit 1, which is mapped to the period-6 orbit $YX = 100101$, which is the period-doubled orbit of the period-3 orbit 100. This explains why the entire diagram can be recognized inside the period-3 window.

The important consequence of this universal behavior is that numerous systems present the same period-doubling cascade leading to a chaotic regime when a parameter varies, as well as the same order of succession of periodic orbit between chaotic regimes, all the more as their return map in a Poincaré section is well approached by a unimodal map. We will give an example of experimental observation of such a route to chaos in a nonlinear chemical system.

In particular, the existence of the period-3 orbit implies the existence of period-n orbits for all the integers n, in fact, of all periodic orbits whose name does not contain 00. Moreover, Li and Yorke in their 1975 paper "Period three implies chaos" demonstrated that the existence of the period-3 orbit also implies the existence of an infinite number of aperiodic orbits (Li and Yorke, 1975).

When the system is not so dissipative but is still associated with a once-folding mechanism in the phase space, much of the universal sequence is preserved for the lowest-period orbits, whereas some higher-period windows do not appear in the same order. This is because the two-dimensional dynamics gives more freedom. In particular, a period-3 orbit no longer implies chaos because such an orbit can live on a torus. In this case, only orbits knotted in some way can be used as a signature of chaos. For more details about this symbolic dynamical approach to the universal sequence, see (Gilmore and Lefranc, 2011).

6.2.5.b The period-doubling cascade and renormalization

Let us now study the universal metric properties of the initial period-doubling cascade. For a large part, they stem from the fact that each period-doubling event looks the same in the vicinity of the bifurcation up to a rescaling and that the period-doubling bifurcation of f where an orbit of period 2^n is created can be viewed as a bifurcation of f^{2^k} where an orbit of period 2^{n-k} is created.

Let us consider a 2^n-periodic orbit in the subharmonic cascade that is stable for $r \in [r_n, r_{n+1}]$. There always exists a parameter $R_n \in [r_n, r_{n+1}]$ for which the critical point $x_c = \frac{1}{2}$ belongs to the orbit. Such an orbit is superstable, its multiplier being zero since $f'(x_c) = 0$. These cycles can be found graphically at the intersections of the bifurcation diagram with the horizontal line $x = x_c = 1/2$ (Fig. 6.22). We denote by d_n the distance between the line $y = 1/2$ and the closest periodic point (Fig. 6.22).

Figure 6.22 illustrates very clearly the self-similarity of the period-doubling cascade, which we already studied in the parameter space, by noting how the distance between two period-doubling bifurcations (the length of the "fork") is approximately a certain fraction $1/\delta$ of the previous value (see Table 6.2). A similar scaling law can be observed in the x space, as we have another universal constant such that

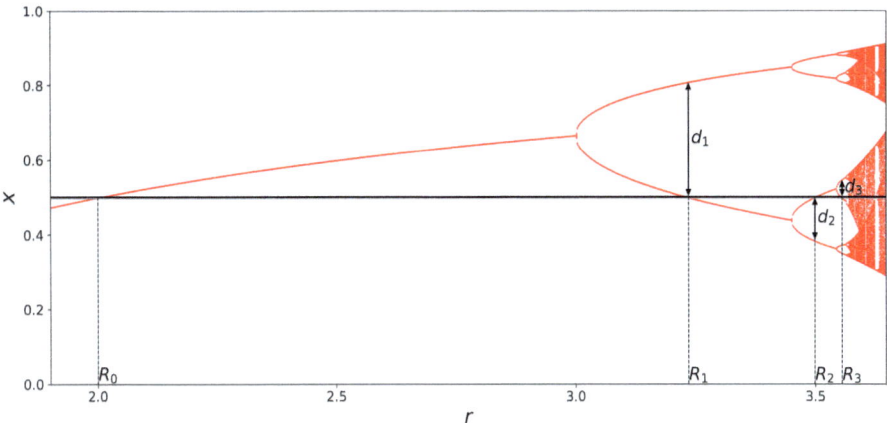

Figure 6.22: Along the period-doubling cascade, there are special parameter values where the critical point belongs to a superstable periodic orbit. At these points the bifurcation diagram intersects the horizontal line $x = x_c$. The distance between the critical point and the closest periodic point at these superstable points obeys a scaling law as described in the text.

$$\frac{d_n}{d_{n+1}} \to -\alpha \quad \text{with } \alpha = 2.5029\ldots.$$

The two exponents α and δ are known as the Feigenbaum constants. Mitchell Feigenbaum (1944–2019) was an American theoretical physicist who realized how to harness the ideas of renormalization such as used in statistical physics to uncover the universality of the period-doubling cascade (Feigenbaum, 1978) (see also (Coullet and Tresser, 1978)) A key idea of renormalization is to rescale a system in such a way that it is indiscernible from the original, allowing us to analyze situations where the system displays some scale invariance, like in second-order phase transitions.

Figures 6.23a–d show the superstable cycles of periods 2^i of $f(x, r)$ at $r = R_i$ for $i = 0, 1, 2, 3$. In each case the critical point $x_c = \frac{1}{2}$ belongs to the cycle, and its distance to the nearest periodic point is what we called d_i. Figures 6.23e–h show the graphs of the $f^{2^i}(x, R_i)$ functions for the same parameters as immediately above. The fixed points of the function $f^{2^i}(x, R_i)$ are the periodic points of periods 2^j of f for $j \leq i$. The stable fixed points correspond to the superstable orbit, whereas the unstable ones correspond to periodic orbits of f of lower period. For example, the unstable fixed point of f^2 in Fig. 6.23f is the period-1 orbit that was superstable in Fig. 6.23a.

In Figs. 6.23e–h, a square delimiting an invariant subinterval is shown. This square is centered on the critical point and one of its edges is located at the unstable fixed point nearest to the critical point. In fact, the latter is the continuation of the periodic point of period 2^{i-1} that was in x_c for $r = R_{i-1}$ and is located approximately half-way between the two points defining the distance d_i, so that the width of the square is approximately d_i.

We see that each square in the bottom row of Fig. 6.23 is a rescaled version of the square in the plot at its left, with a scale factor approximately equal to what we de-

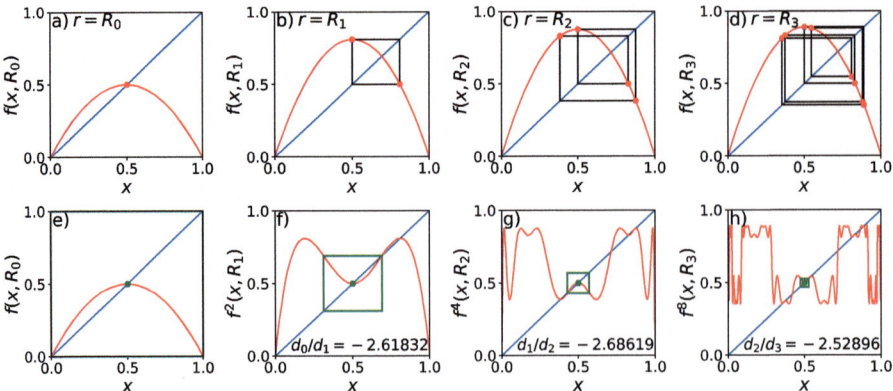

Figure 6.23: Top row: graph of $f(x, R_i)$ with a superstable cycle of period 2^i at $r = R_i$ for (a) $i = 0$; (b) $i = 1$; (c) $i = 2$; (d) $i = 3$. In each case the critical point at $x_c = \frac{1}{2}$ belongs to the cycle. Bottom row: graph of $f^{2^i}(x, R_i)$ together with a square delimiting an invariant subinterval surrounding the critical point for (e) $i = 0$; (f) $i = 1$; (g) $i = 2$; (h) $i = 3$. The boundary of the square at $r = R_i$ corresponds to the new position of the fixed point coinciding with the critical point for $r = R_{i-1}$. The sizes of the successive squares are approximately in the ratio $-\alpha$.

fine as $-\alpha$, where the minus sign takes into account that the graph is flipped upside-down.

This construction can be repeated at higher orders, leading to the same observation that the graph of the function f^{2^n} around $x_c = 1/2$ is similar to that of the function $f^{2^{n-1}}$, but reduced by a factor $-\alpha$. Denoting $\tilde{f}(x) = f(x_c + x)$, where we have shifted the horizontal axis so that $\tilde{f}$ has its maximum at the origin, we have

$$\tilde{f}^{2^{n-1}}(x, R_{n-1}) \simeq -\alpha \tilde{f}^{2^n}\left(-\frac{x}{\alpha}, R_n\right), \tag{6.16}$$

from which we deduce that

$$\tilde{f}(x, R_0) \simeq (-\alpha)^n \tilde{f}^{2^n}\left(\frac{x}{(-\alpha)^n}, R_n\right). \tag{6.17}$$

As $n \to \infty$, the right-hand term of (6.17) converges to a universal one-hump function g_0, which is independent of f. Since we probe an infinitesimal neighborhood of the maximum of the function, all functions with a quadratic maximum look the same.

Starting from a higher-period superstable point, we similarly have

$$\tilde{f}(x, R_i) \simeq (-\alpha)^n \tilde{f}^{2^n}\left(\frac{x}{(-\alpha)^n}, R_{n+i}\right). \tag{6.18}$$

Again, the right-hand term of (6.18) converges to a universal function that has a superstable 2^i cycle as $n \to \infty$. If we now take $i \to \infty$ so that the two sides of (6.18) are evaluated at R_∞, then this leads to a functional equation defining a universal function with a superstable 2^∞ cycle that is independent of the function f studied and describes the structure of the invariant set at the onset of chaos:

$$g(x) = -\alpha g\left[g\left(-\frac{x}{\alpha}\right)\right].$$

(6.19)

Equation (6.19) expresses the fact that the function g is invariant if we apply to it the transformation given in the right-hand side, where we rescale x, iterate f, and rescale the result. This equation has only a solution if

$$\alpha = 2.502907875\ldots,$$

which can be found by writing a Taylor expansion of g and solving the equations for the coefficients. We can always choose $g(0) = 1$ and $g'(0) = 0$, since we have shifted the origin to the location of the maximum. We can see that then (6.19) implies that $\alpha = -1/g(1)$. The curve in Figure 6.23h gives a simplified idea of what $g(x)$ looks like.

6.2.5.c Universal period-doubling route to chaos

It is now time to come back to experiments. We saw in Chapter 5 that in very dissipative systems, it is common to observe a unidimensional first return map and that this map contains the two central mechanisms generating chaos, stretching and squeezing, which translate in a bell-shape function.

The following example is based on the Belousov–Zhabotinskii reaction studied by Simoyi et al. (Simoyi et al., 1982). The system implies about 25 chemical species in a stirred reactor fed continuously with different reactants. The dynamics of the concentration of one of the reactants, the bromide ion, is measured with a specific probe. Several time series of this probe are reproduced in Fig. 6.24, each corresponding to a given flow rate.

The authors clearly observed the first steps of a period-doubling cascade when varying the flow rate: the periodic regime at period T changes to a $2T$ regime and then to a $4T$ regime (top row of Fig. 6.24). As the control parameter varies, different periodic regimes are observed beyond the onset of chaos, corresponding to periodic windows, in the order predicted by the universal sequence. In particular, the bottom row of Fig. 6.24 shows a $3T$ periodic regime, followed by a $6T$ oscillation, which is the first step of the period-doubling cascade embedded inside the $3T$ window.

Figure 6.25(a) shows a chaotic attractor reconstructed from the time series using the time-delay method. In spite of the large number of species implied in the reaction, the attractor can be embedded in a three-dimensional phase space, and a Poincaré map can be performed using a plane orthogonal to the figure, leading to a one-dimensional first return map (Fig 6.25(b)). This first-return map is unimodal and thus belongs to the universal class described by the logistic function.

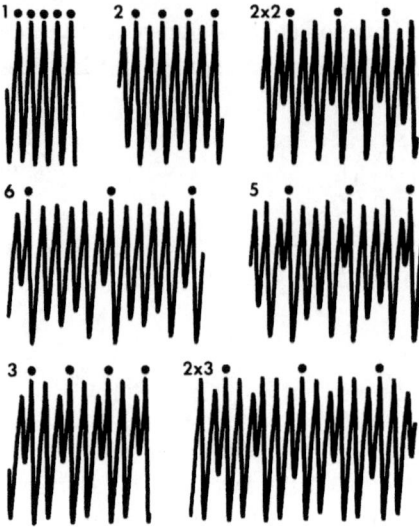

Figure 6.24: Reprinted figure with permission from [R. H. Simoyi, *et al.*, *Phys. Rev. Lett.* **49**, 245 (1982)] Copyright (1982) by the American Physical Society. Concentration of the bromide ion as a function of time. The dots indicate the beginning of the repeated pattern.

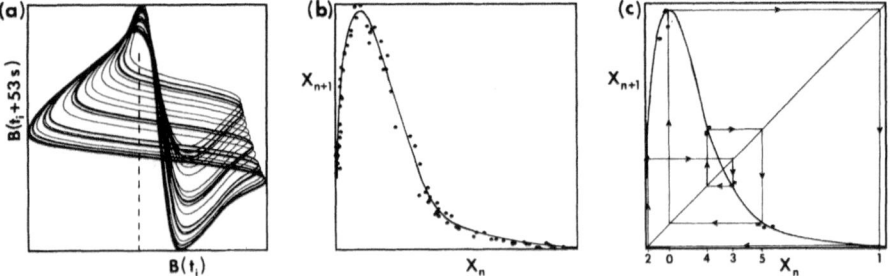

Figure 6.25: Reprinted figure with permission from [R. H. Simoyi, *et al.*, *Phys. Rev. Lett.* **49**, 245 (1982)] Copyright (1982) by the American Physical Society (a) 2D projection of the attractor reconstructed from a chaotic time series using the time delay method. A Poincaré section is performed using a plane orthogonal to the figure, as indicated by the dashed line. (b) First-return map constructed from the Poincaré section (a). (c) As the chaotic regime analyzed is very close to a period-6 window, the location of the corresponding cycle can be approximated using the continuous function fitting the first-return map in (b).

The period-doubling route to chaos has been observed in numerous experimental systems in the 1980s ranging from fluid mechanics (Giglio et al., 1981; Libchaber et al., 1982) to electric circuits (Linsay, 1981; Testa et al., 1982) or lasers (Arecchi et al., 1982; Midavaine et al., 1985). In particular, lasers are an interesting system to explore chaos thanks to their strong nonlinearity, fast time scales, and good signal-to-noise ratio. Figure 6.26 shows a bifurcation diagram obtained in real time with a CO_2 laser with modulated losses.

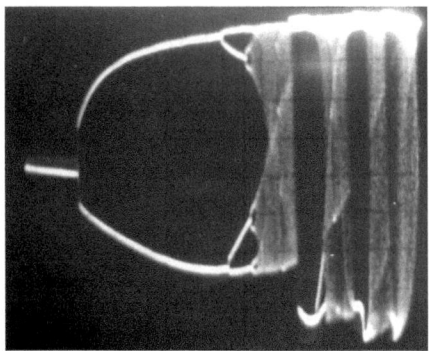

Figure 6.26: Real-time bifurcation diagram observed with a CO_2 laser with losses modulated at a frequency of about 400 kHz. A sample-and-hold module is synchronized with the driving signal to obtain one laser intensity sample per modulation period. The amplitude of the driving is then slowly swept. The horizontal coordinate is the modulation amplitude, whereas successive samples measured for this amplitude are displayed vertically. The structure of the diagram is identical to that of the logistic map: the same periodic windows are observed in the same order. However, the main periodic windows are wider, which can be related to a different stability of the orbit delimiting the boundary of the attraction basin.

6.3 Conclusions

In this last chapter, we first described quantitative tools to characterize strange attractors, Lyapunov exponents and fractal dimension. These measures of chaos quantify the essential properties of chaos: sensitivity to initial conditions, which is due to stretching, and the fractal structure of strange attractors, which is due to folding and squeezing. We also noted that these two geometrical processes can be identified using a topological analysis of unstable periodic orbits.

Then we detailed an ubiquitous route to chaos, the subharmonic cascade, which is remarkable by its universal properties. Although very common, it is not the only scenario. In particular, another mechanism leads from quasi-periodicity to chaos through the transformation of an invariant torus into a fractal object (Ruelle and Takens, 1971).

In this short introduction to nonlinear dynamics, we did not try to be exhaustive but rather to lay down the fundamental concepts of this fascinating field of science and to give the reader the desire to learn more. For this, we refer you to the bibliography at the end of this textbook, where you will find references of excellent and more advanced books on nonlinear dynamics and chaos.

Exercises

Sierpiński carpet

Consider the object built recursively in the following way (Fig. 6.27): a square of unit side is divided into nine smaller squares of equal dimensions, and the central square is removed (step 1). We repeat this operation on each of the 8 remaining squares (step 2) and then again recursively an infinite number of time. The object obtained for an infinite number of iterations is fractal and is called the Sierpiński carpet.

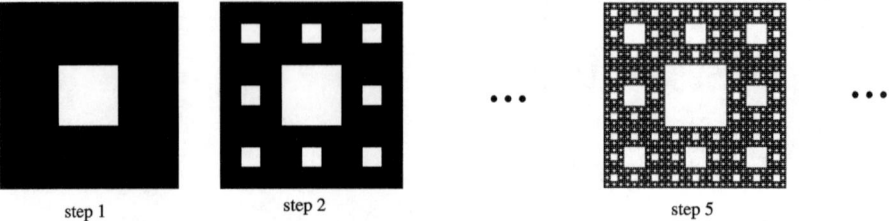

step 1 step 2 step 5

Figure 6.27: First steps of the construction of the Sierpiński carpet.

1. Compute the area of the object at the nth iteration.
2. Show that the area of the Sierpiński carpet is zero.
3. Compute the box counting dimension of the Sierpiński carpet.

Numerical study of the Lorenz attractor

Consider the Lorenz system

$$\dot{X} = \mathrm{Pr}(Y - X),$$
$$\dot{Y} = -XZ + rX - Y,$$
$$\dot{Z} = XY - bZ,$$

with $\mathrm{Pr} = 10$, $b = 8/3$, and $r = 28$.

1. Apply the method described in Section 6.1.1.a (Fig. 6.2) to obtain an estimation of the largest Lyapunov exponent of the Lorenz attractor: define an initial condition from a perturbation of a point of the attractor and compute the distance between the trajectories as a function of time. Plot the evolution of the distance between the trajectories with time in a semilogarithmic graph and use a fit of the linear part of the graph to deduce the Lyapunov exponent.
2. Use the correlation method (Section 6.1.2.d) to compute the fractal dimension of the Lorenz attractor.

The "tent" map

Consider the one-dimensional function defined as follows:

$$f(x) = rx \quad \text{for } 0 \le x < 0.5,$$
$$f(x) = r - rx \quad \text{for } 0.5 < x \le 1.$$

This function is defined for $x \in [0,1]$. The parameter r takes values between 0 and 2. We are interested in the iterates of this function, i. e., the discrete dynamics of the form

$$x_{n+1} = f(x_n).$$

1. Draw the function to study for $r = 0.5$, $r = 1$, and $r = 2$. Why do we restrict $r \le 2$?
2. For $r \in]0,1[$, calculate the fixed points of the function and their stability.
3. The same question for $r \in]1,2]$.
4. For $r \in]1,2]$, compute the image by f of the following intervals:

$$0 \le x \le \frac{1}{2r},$$
$$\frac{1}{2r} \le x \le 0.5,$$
$$0.5 \le x \le 1 - \frac{1}{2r},$$
$$1 - \frac{1}{2r} \le x \le 1.$$

5. Deduce the expression of $f^2(x)$ in each of those four intervals.
6. Draw the graph of $f^2(x)$ for $r = 2$, $r = 1.5$, and $r \ge 1$.
7. Using the previous questions, determine the fixed points of $f^2(x)$ and their stability.
8. Consider an N-cycle: $x_0, x_1, \ldots, x_{N-1}$ and the function $g = f^N$. Show that for any x_i,

$$g'(x_i) = \prod_{i=0}^{N-1} f'(x_i).$$

9. Does f have stable limit cycles for $r \in]1,2]$?
10. What is the Lyapunov exponent of f? For what values of r is the dynamics chaotic? Is your result in accordance with the previous questions?
11. Using $r = 2$, describe the stretching and squeezing mechanism at play on the interval $[0,1]$.

Subharmonic cascade in the Rössler model

Consider the Rössler system

$$\dot{X} = -Y - Z,$$
$$\dot{Y} = X + aY,$$

$$\dot{Z} = b + Z(X - c),$$

with $a = b = 0.2$ and $c \in [2.5, 6]$.

1. Integrate numerically the system and observe the asymptotic dynamics for $c = 2.5$, 3.5, 4, 5, 5.2.
2. Write a function selecting the local maximums of X in the asymptotic regime and plot their values as a function of c to obtain a bifurcation diagram.
3. Draw the first return map using the method of Section 5.4.2, i. e., by plotting the local maximum X_{i+1} as a function of the previous one X_i.

Topological entropy, metric entropy, and Lyapunov exponent in the logistic map

Here we make use of the symbolic coding defined for the logistic map in Section 6.2.1 to compute the invariants of the dynamics. First, choose an integer l that will be the maximum length of symbolic sequences

1. For a number of parameters $0 \le r_i \le 4$, compute the probability p_i of occurrence of each possible finite symbol sequence of length l. Given an initial condition x_0, this is done by computing the symbol sequence of length l associated with $x_0, x_1, \ldots, x_{l-1} = f^{l-1}(x_0)$. Then compute the subsequent iterates, each time discarding the leading symbol and adding the new symbol at the end. How do you make sure that the probability estimate has reasonably converged?
2. Once the p_i have been measured for all sequences of length l, compute the following quantities and plot them in the same graph:
   - The Lyapunov exponent, as defined in (6.4).
   - The *metric entropy* $h = -\frac{1}{l} \sum_i p_i \ln p_i$ (where we can recognize the Shannon entropy per symbol, representing the amount of information gained at each iteration).
   - The topological entropy $h_T = \frac{\ln \mathcal{N}(l)}{l}$, where $\mathcal{N}(l)$ is the number of different symbolic sequences of length l observed (i. e., which have a nonzero probability of occurrence).
3. Check your results for different values of l. The smaller the l, the less precise the result. The larger the l, the longer the computation if one requires that each possible sequence has been sampled enough. Find a good compromise.
4. Are your findings consistent with the following facts?
   - The metric entropy is equal to the Lyapunov exponent. For higher-dimensional systems, it is conjectured that $h_T = \sum_{\lambda_i > 0} \lambda_i$, the sum of the positive Lyapunov exponents.
   - The topological entropy is an upper bound of the metric entropy. Note that the topological entropy can be recovered from the metric entropy assuming that all p_i are equal, which is known to maximize the entropy.

Note that the topological entropy computed as indicated above drops to a small value in each periodic window, since we only observe the periodic sequence. Actually, it should be computed taking also into account all the unstable orbits that are no longer visited. We would then find that the real topological entropy increases smoothly with parameter r.

Universal sequence

1. Show that the beginning of the period-3 window corresponds to the appearance of two orbits with sequence 101, one of which becomes 100 after becoming superstable.
2. Show that any periodic sequence not containing 00 precedes 101 according to order (6.14). Conclude that when the period-3 window begins, infinitely many periodic orbits have already been created, hence "Period-3 implies chaos" (Li and Yorke, 1975).
3. Show that for the last window of period p, the orbit that undergoes period-doubling has symbolic name 10^{p-1} (that is, its rightmost point is at the right of any other rightmost point of the same period).
4. Show that all orbits whose names contain 0^{n-1} but not 0^n appear before the orbits whose names contain 0^n but not 0^{n+1}. Interestingly, this rule completely breaks down in once-folding two-dimensional maps, whereas most of the universal sequence remains true for lowest-period orbits.

For further reading

– Abraham R. H. and Shaw C. D. (1992). Dynamics: The Geometry of Behavior, 2nd Edition. Basic Books.
– Alligood K. T., Sauer T. D. and Yorke J. A. (2000). Chaos: An Introduction to Dynamical Systems. Springer.
– Bergé P., Pomeau Y. and Vidal C. (1987). Order within Chaos. Wiley-VCH.
– Cvitanović P., Artuso R., Mainieri R., Tanner G. and Vattay G. (2020). Chaos: Classical and Quantum. https://ChaosBook.org. Niels Bohr Institute, Copenhagen.
– Devaney R. L. (1989). An Introduction to Chaotic Dynamical Systems, 2nd Edition. Addison-Wesley.
– Gilmore R. and Lefranc M. (2011). The Topology of Chaos: Alice in Stretch and Squeezeland, 2nd Edition. Wiley-VCH.
– Gilmore R. and Letellier C. (2007). The Symmetry of Chaos. Oxford University Press.
– Guckenheimer J. and Holmes P. (2002). Nonlinear Oscillations, Dynamical Systems and Bifurcations of Vector Fields. Springer.
– Hirsch M. W., Smale S. and Devaney R. L. (2012). Differential Equations, Dynamical Systems and an Introduction to Chaos, 3rd Edition. Academic Press Inc.
– Katok A. and Hasselblatt B. (1995). Introduction to the Modern Theory of Dynamical Systems. Cambridge University Press.
– Jackson E. A. (1992). Perspectives of Nonlinear Dynamics 1&2. Cambridge University Press.
– Manneville P. (2010). Instabilities, Chaos and Turbulence. Imperial College Press.
– Ott E. (2002). Chaos in Dynamical Systems, 2nd Edition. Cambridge: Cambridge University Press.
– Strogatz S. H. (2015). Nonlinear Dynamics and Chaos: With Applications to Physics, Biology, Chemistry, and Engineering, 2nd Edition. CRC Press.
– Tufillaro N. B., Abbott T. and Reilly J. (1992). An Experimental Approach to Nonlinear Dynamics and Chaos. Basic Books.
– Wiggins S. (2003). Introduction to Applied Nonlinear Dynamical Systems and Chaos, 2nd Edition. Springer.

https://doi.org/10.1515/9783110677874-007

Bibliography

Abraham, R. and Shaw, C. (1992). *Dynamics-The geometry of behavior*. Addison-Wesley.

Arecchi, F. T., Meucci, R., Puccioni, G., and Tredicce, J. (1982). Experimental evidence of subharmonic bifurcations, multistability, and turbulence in a q-switched gas laser. *Phys. Rev. Lett.*, 49:1217–1220.

Arnold, V. (1965). Small denominators. i. mappings of the circumference onto itself. *Am. Math. Soc. Transl. (2)*, 46:213–284.

Arnold, V. I., Gusein-Zade, S. M., and Garchenko, A. N. (2012). *Singularities of differentiable maps, Volume 1*. Birkhäuser.

Barber, D. J. and Loudon, R. (1989). *An introduction to the properties of condensed matter*. Cambridge University Press.

Coullet, P. and Tresser, C. (1978). Itération d'endomorphismes et groupe de renormalisation. *J. Phys., Colloq. C5*, 25–28.

Ecke, R. E., Farmer, J. D., and Umberger, D. K. (1989). Scaling of the arnold tongues. *Nonlinearity*, 2(2):175.

Eckmann, J. P., Kamphorst, S. O., Ruelle, D., and Ciliberto, S. (1986). Liapunov exponents from time series. *Phys. Rev. A*, 34:4971–4979.

Feigenbaum, M. (1978). Quantitative universality for a class of nonlinear transformations. *J. Stat. Phys.*, 19:25–52.

Feigenbaum, M. J. (1980). Universal behavior in nonlinear systems. In Cvitanović, P., editor, *Universality in chaos*, pages 49–84. Taylor & Francis.

Flaherty, J. E. and Hoppensteadt, F. (1978). Frequency entrainment of a forced van der pol oscillator. *Stud. Appl. Math.*, 58(1):5–15.

Giglio, M., Musazzi, S., and Perini, U. (1981). Transition to chaotic behavior via a reproducible sequence of period-doubling bifurcations. *Phys. Rev. Lett.*, 47:243–246.

Gilmore, R. (1981). *Catastrophe theory for scientists and engineers*. John Wiley & Sons.

Gilmore, R. and Lefranc, M. (2011). *The topology of chaos: Alice in stretch and squeezeland, 2nd Edition*. Wiley-VCH.

Ginoux, J. and Letellier, C. (2012). Van de pol and the history of relaxation oscillations: towards the emergence of a concept. *Chaos*, 22:023120.

Grassberger, P. (1983). Generalized dimensions of strange attractors. *Phys. Lett. A*, 97(6):227–230.

Grassberger, P. and Procaccia, I. (1983). Measuring the strangeness of strange attractors. *Phys. D, Nonlinear Phenom.*, 9(1):189–208.

Holmes, P. and Rand, D. (1976). The bifurcations of duffing's equation: An application of catastrophe theory. *J. Sound Vib.*, 44(2):237–253.

Hudson, J. L. and Mankin, J. C. (1981). Chaos in the belousov-zhabotinskii reaction. *J. Chem. Phys.*, 74:6171.

Hénon, M. (1976). A two-dimensional mapping with a strange attractor. *Commun. Math. Phys.*, 50.

Jensen, M. H., Bak, P., and Bohr, T. (1984). Transition to chaos by interaction of resonances in dissipative systems. i. circle maps. *Phys. Rev. A*, 30:1960.

Kaplan, J. L. and Yorke, J. A. (1979). In Peitgen, H.-O. and Walther, H.-O., editors, *Functional differential equations and approximations of fixed points: Proceedings, Bonn, July 1978*, page 204, Berlin. Springer-Verlag.

Katok, A. and Hasselblatt, B. (1995). *Introduction to the modern theory of dynamical systems*. Cambridge University Press.

Laroche, C., Labbé, R., Pétrélis, F., and Fauve, S. (2012). Chaotic motors. *Am. J. Phys.*, 80(2):113–121.

Lefranc, M. and Glorieux, P. (1993). Topological analysis of chaotic signals from a CO_2 laser with modulated losses. *Int. J. Bifurc. Chaos Appl. Sci. Eng.*, 3:643–649.

Li, T.-Y. and Yorke, J. A. (1975). Period three implies chaos. *Am. Math. Mon.*, 82(10):985–992.

Libchaber, A., Laroche, C., and Fauve, S. (1982). Period doubling cascade in mercury, a quantitative measurement. *J. Phys. Lett.*, 43(7):211–216.

https://doi.org/10.1515/9783110677874-008

Linsay, P. S. (1981). Period doubling and chaotic behavior in a driven anharmonic oscillator. *Phys. Rev. Lett.*, 47:1349–1352.

Lorenz, E. N. (1963). Deterministic nonperiodic flow. *J. Atmos. Sci.*, 20(2):130–141.

Midavaine, T., Dangoisse, D., and Glorieux, P. (1985). Observation of chaos in a frequency-modulated CO_2 laser. *Phys. Rev. Lett.*, 55:1989–1992.

Pfeuty, B., Thommen, Q., and Lefranc, M. (2011). Robust entrainment of circadian oscillators requires specific phase response curves. *Biophys. J.*, 100(11):2557–2565.

Press, W. H., Teukolsky, S. A., Vetterling, W. T., and Flannery, B. P. (2007). *Numerical recipes: The art of scientific computing*. Cambridge University Press.

Prigogine, I. and Lefever, R. (1968). Symmetry breaking instabilities in dissipative systems. ii. *J. Chem. Phys.*, 48(4):1695–1700.

Rössler, O. E. (1976). An equation for continuous chaos. *Phys. Lett. A*, 57(5):397–398.

Ruelle, D. (1980). Strange attractors. *Math. Intell.*, 2:126–137.

Ruelle, D. and Takens, F. (1971). On the nature of turbulence. *Commun. Math. Phys.*, 20:167–192.

Rényi, A. (1959). On the dimension and entropy of probability distributions. *Acta Math. Acad. Sci. Hung.*, 10:193–215.

Sel'Kov, E. (1968). Self-oscillations in glycolysis 1. a simple kinetic model. *Eur. J. Biochem.*, 4(1):79–86.

Simoyi, R. H., Wolf, A., and Swinney, H. L. (1982). One-dimensional dynamics in a multicomponent chemical reaction. *Phys. Rev. Lett.*, 49:245–248.

Smale, S. (1967). Differentiable dynamical systems. *Bull. Am. Math. Soc.*, 73:747–817.

Strogatz, S. H. (2018). *Nonlinear dynamics and chaos: With applications to physics, biology, chemistry, and engineering*. CRC Press.

Testa, J., Pérez, J., and Jeffries, C. (1982). Evidence for universal chaotic behavior of a driven nonlinear oscillator. *Phys. Rev. Lett.*, 48:714–717.

Tyson, J. J. (1991). Modeling the cell division cycle: cdc2 and cyclin interactions. *Proc. Natl. Acad. Sci.*, 88(16):7328–7332.

Index

https://doi.org/10.1515/9783110677874-009